AF478153

Developing Agricultural Trade

The Role of Government in Adjusting Economies

General Editor: **Richard Batley, International Development Department, School of Public Policy, University of Birmingham**

Over the last two decades there has been a strong emphasis on reducing the role of government and on reforming traditional public sector bureaucracies. The new conventional view has become that, where possible, services should not be provided directly by government but be contracted out or privatized. Where this is not possible, the predominant view has been that the public sector itself should change by setting up semi-autonomous agencies and by making public management more performance- and customer-oriented.

This series investigates the application of such reforms in Africa, Asia and Latin America. Underlying the enquiry is the question whether reforms which were initially conceived in countries such as Britain and New Zealand are appropriate in other contexts. How much sense do they make where levels of public management capacity, market development, resources, political inclusiveness, legal effectiveness, political and public economic stability are quite different?

To investigate these issues, the series covers four service sectors selected to be representative of types of public sector activity – health care, urban water supply, agricultural marketing services and business development services.

Titles include:

Michael Hubbard
DEVELOPING AGRICULTURAL TRADE
New Roles for Government in Poor Countries

Paul Jackson
BUSINESS DEVELOPMENT IN ASIA AND AFRICA
The Role of Government Agencies

Anne Mills, Sara Bennett and Steven Russell
THE CHALLENGE OF HEALTH SECTOR REFORM
What Must Governments Do?

The Role of Government in Adjusting Economies
Series Standing Order ISBN 0–333–94618–9
(*outside North America only*)

You can receive future titles in this series as they are published by placing a standing order. Please contact your bookseller or, in case of difficulty, write to us at the address below with your name and address, the title of the series and the ISBN quoted above.

Customer Services Department, Macmillan Distribution Ltd, Houndmills, Basingstoke, Hampshire RG21 6XS, England

Developing Agricultural Trade

New roles for government in poor countries

Michael Hubbard
International Development Department
University of Birmingham, UK

With

Marisol Smith, Frank Ellis, Gideon Onumah,
Andrew Shepherd, Peter Lewa and Renu Kohli

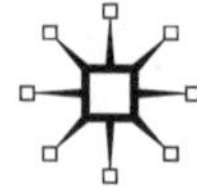

First published 2003 by
PALGRAVE MACMILLAN
Houndmills, Basingstoke, Hampshire RG21 6XS and
175 Fifth Avenue, New York, N.Y. 10010
Companies and representatives throughout the world

PALGRAVE MACMILLAN is the global academic imprint of the Palgrave
Macmillan division of St. Martin's Press, LLC and of Palgrave Macmillan Ltd.
Macmillan® is a registered trademark in the United States, United Kingdom
and other countries. Palgrave is a registered trademark in the European
Union and other countries.

ISBN 0–333–73619–2

This book is printed on paper suitable for recycling and made from fully
managed and sustained forest sources.

A catalogue record for this book is available from the British Library.

Library of Congress Cataloging-in-Publication Data

Hubbard, Michael.
    Developing agricultural trade : new roles for government in poor
    countries / Michael Hubbard.
       p.   cm.—(The role of government in adjusting economies)
    Includes bibliographical references and index.
    ISBN 0–333–73619–2 (cloth)
1. Produce trade—Government policy—Developing countries.   2. Agriculture
and state—Developing countries.   I. Title.   II. Series.

HD9018.D44 H83 2002
380.1′41′091724—dc21

                                                              2002028750

10   9   8   7   6   5   4   3   2   1
12   11   10   09   08   07   06   05   04   03

Printed and bound in Great Britian by
Antony Rowe Ltd, Chippenham and Eastbourne

Contents

List of Figures

List of Tables

Preface

Governments worldwide have intervened heavily in agricultural trade to protect their food supplies, farmers and exports. Many are now seeking to liberalise domestic and external trade and develop the contribution of private trade to food security, price stabilisation and export growth. This book is concerned with how governments are managing trade liberalisation and reform of public services to agricultural trade. The focus is on developing countries, with case studies from India, Sri Lanka, Ghana, Zimbabwe, Kenya and Côte d'Ivoire. The conclusions highlight the individuality of approaches, the conflicts involved and the varying capability and will of governments to take on new roles.

The book originates in a collaborative, multi-sectoral research project on the role of government in adjusting economies funded by the Economic and Social Committee on Research (ESCOR) of the Department for International Development (DFID).[1] The study on the changing role of government in agricultural trade involved collaboration of the International Development Department, University of Birmingham and the Overseas Development Group, University of East Anglia, with local consultants in Côte d'Ivoire, Ghana, India, Kenya, Sri Lanka and Zimbabwe. The study was coordinated by Michael Hubbard of the University of Birmingham, with Marisol Smith managing the contributions from the University of East Anglia.

Extracts from papers produced by the study have been adapted and updated to provide most of the case material included in this book. The papers are the following, all published by the International Development Department, School of Public Policy, University of Birmingham:

Hubbard, M. and Smith, M. (1996) 'Agricultural marketing sector review'. Paper 6.

Shepherd, A. and Onumah, G. (1997) 'Liberalised agricultural markets in Ghana: the roles and capacity of government'. Paper 12.

Smith, M. and Ellis, F. (1997) 'The role of the state in agricultural marketing in Sri Lanka'. The Role of Government in Adjusting Economies. Paper 19.

Kohli, R. and Smith, M. (1998) 'The role of the state and agricultural marketing reform in India'. The Role of Government in Adjusting Economies. Paper 29.

Atse, D. (1999) 'Reforming the cocoa sector in Côte d'Ivoire: the case of the electronic trading system'. The Role of Government in Adjusting Economies. Paper 34.

Chikandi, S. (1999) 'ZFU's pilot market information service: an analysis of the service and early consumer reaction to it', in M. Hubbard (ed.) 'The impact of liberalisation on public services to agricultural marketing: case studies of change options in Zimbabwe'. The Role of Government in Adjusting Economies. Paper 36.

Rusike, J. (1999) 'Managing the public role in liberalised seed markets in Zimbabwe', in M. Hubbard (ed.), 'The impact of liberalisation on public services to agricultural marketing: case studies of change options in Zimbabwe'. The Role of Government in Adjusting Economies. Paper 36.

Sukume, C. (1999) 'Quality assurance in liberalised meat markets', in M. Hubbard (ed.), 'The impact of liberalisation on public services to agricultural marketing: case studies of change options in Zimbabwe'. The Role of Government in Adjusting Economics. Paper 36.

Hubbard, M. (1999) 'Governing open agricultural markets in Zimbabwe: state capacity and performance'. The Role of Government in Adjusting Economies. Paper 37.

Acknowledgement is due to the many who facilitated fieldwork and participated in the intellectual process of which this book is the result. Particular mention must be made of the contribution of the advisory panel for the study: Martin Hebblethwaite of the Natural Resources Institute; John Marsh of the Centre for Agricultural Strategy, University of Reading; and Barbara Harriss-White, University of Oxford. The International Development Department, University of Birmingham, provided a sabbatical for the preparation of this book.

Michael Hubbard

List of Abbreviations

Abbreviations have been avoided where possible. The main ones used are the following:

APL	'above poverty line'. Households earning incomes exceeding poverty line in India
BIS	Bureau of Indian Standards
BPL	'below poverty line'. Households earning incomes below poverty line in India
CFU	Commercial Farmers Union, Zimbabwe
CMC	Cocoa Marketing Company, Ghana
COTCO	Cotton Marketing Company, Zimbabwe. Successor to Cotton Marketing Board
CSC	Cold Storage Company, Zimbabwe. State meat processing company, successor to Cold Storage Commission
CSRP	Cereals Sector Reform Programme, Kenya
CWC	Central Warehousing Corporation, India. Warehousing company owned by federal government
CWE	Cooperative Welfare Establishment, Sri Lanka
DFR	Department of Feeder Roads, Ghana
DZL	Dairibord Zimbabwe Ltd, successor to Dairy Marketing Board
ECA	Essential Commodities Act, India
ERP	Economic Recovery Plan, Ghana
ESAP	Economic and Structural Adjustment Programme, Zimbabwe
FCD	Food Commissioner's Department, Sri Lanka
FCI	Food Corporation of India
GFDC	Ghana Food Distribution Corporation
GMB	Grain Marketing Board, Zimbabwe
GPS	Guaranteed Price Scheme, Sri Lanka
HACCP	Hazardous Action Critical Control Points. Management system for food safety
LBC	Licensed Buying Company. Private companies allowed to buy cocoa from farmers, Ghana
MOLA	Ministry of Lands and Agriculture, Zimbabwe
NCPB	National Cereals and Produce Board, Kenya

NIE	New Institutional Economics
NPM	New Public Management
PBC	Produce Buying Company, state-owned company buying cocoa from farmers, Ghana
PDS	Public Distribution System, India
PMB	Paddy Marketing Board, Sri Lanka
QCD	Quality Control Division of Cocobod, Ghana
SAFEX	South African Futures Exchange
SWC	State Warehousing Corporation, India. State level public warehousing companies
ZFU	Zimbabwe Farmers Union. For small farmers
ZIMACE	Zimbabwe Agricultural Commodities Exchange

Part I
Introduction

1
Government and Markets: Theory and Concepts

The middle decades of the twentieth century saw a rise in the role of the state worldwide. State expenditure as a proportion of GDP rose dramatically, influenced in part by the replacement of the market with the state in the Soviet Union and China to stem perceived evils of capitalism, and by the popular clamour post-World War Two for better health, education and infrastructure. But with market economies outperforming centrally planned economies, and much state enterprise accumulating losses, the tide of ideology and state reform in the 1980s and 1990s turned towards expanding the private sector. The main targets for privatisation were state manufacturing and state farming enterprises. But state services were brought into the privatisation focus too.

A new image of the desirable public role emerged. The ideal was the 'enabling state', which carries out only those tasks which the private sector cannot, supports market development, and balances its budget (in an average year) so that currency stability is underpinned and private savings are available for private investment.

Taking steps to realise this ideal has principally involved liberalisation (reducing controls on trade, domestic and external) and privatisation (state withdrawing from goods and services which can be provided as well or better by private firms). Both were widely studied in the 1980s and 1990s.

Less widely studied, particularly in developing countries, have been the implications of liberalisation and privatisation for the organisation and performance of the state itself. Broadly, the shift towards open markets and the resulting rise in private sector services requires a shift in public service provision out of services competing with the private sector, and towards a range of activities which support market development (enabling activities). Understanding how governments in developing countries make or fail to make this adjustment in public services is the purpose of this book, based on country case studies in Africa and Asia.

The dominant model for public sector reform since the 1980s has been that of 'new public management' (NPM). In the NPM model the market is used for the delivery of public services. The state confines itself to policy, planning and financing services, while competition among service providers bidding for state contracts to deliver public services is the basis for raising performance in services.

The research project of which this book is the outcome focused on changing modes of provision of public services. It was motivated by the concern that NPM was being advocated too unquestioningly in developing countries, perhaps because it appears to be a means of providing public services even where the state is weak or not oriented to providing good services.

The underlying hypotheses of the research were that:

1 NPM is not always the best means for raising performance of public services – it depends on the institutional environment.
2 NPM is not a quick solution where the state machinery has decayed.

This chapter provides theoretical background and sets out the approach of this book. Since the concern of this book is with adjustment by governments of their roles in agricultural markets, we begin with the theory of government's role in markets, discussing market failure, state failure, bureaucracy and institutions. We then turn to the tasks for government in commercialising public services. Finally, the approach taken in this book is outlined.

The theory of government's role in markets

The study makes use of three strands of theory concerning government's role in markets, namely public goods, institutions and state failure.

Public goods

There is a range of economic arguments for state intervention to ensure goods and services are produced and consumed in sufficient quality and quantity. Economic theory identifies 'pure' public goods by the technical characteristic of non-subtractability (no one's consumption of the good reduces consumption by anyone else). For example, street lighting, national defence and public administration are in their nature consumed by all, and are therefore goods which no individual will have an incentive to invest in, if hoping to reap returns by selling them.

Beyond this narrow technical characteristic, defining goods as 'public' is arguably cultural (Malkin and Wildavsky 1991, quoted by Walsh 1995: 7) or circumstantial – resulting from the limitations of markets (market failure), and the technology, institutions (the social 'rules of the game') and infrastructure underlying markets at a particular time and place. The resulting surface characteristics by which market failure is usually identified include: monopoly, externalities, merit goods, information asymmetries and absence of agreed or enforceable property rights over the good or its means of production (non-excludability) – all of which result in production which is insufficient in quantity or quality. Developments in technology, institutions and infrastructure bring about market development, reduce market failure, and thereby reduce 'public goods' content defined by these characteristics.

More broadly, it can be argued that the interdependence of modern society makes maintaining stability and confidence in the social system a key public good:

> The more complex society becomes, and more the actions of one person are intricately bound up with those of another, the more likely it is that any good will have public good characteristics...At the very least, public services may be seen as creating legitimacy for the social system or preventing unrest, which are public goods. (Walsh 1995: 7)

Where the good is an essential (e.g. food, fuel, water) and disruption of supplies likely to be worsened by hoarding resulting in social unrest, public intervention might therefore be expected.

Second, and against economic theory,[1] is the political argument that the state should intervene in markets to alter who benefits from them. In its most extreme form, this motive was used to justify abolishing markets altogether under communist central planning. Less extremely, it has been used to justify extensive control of markets – as in the case of food marketing (see Chapter 2).

Market failure analysis is necessarily broad. However, it provides a rough indication of 'public goods' content and whether the basis for public intervention is technical, or related to underdevelopment of the market and feared instability, or politically motivated by redistribution objectives.[2]

Institutions

The quality of institutions (laws, rules, conventions, property rights) limits market development and underlies the relation between state and markets. Institutions are deeply entrenched in the development path of that society ('path dependence'). If the rules continually block market development which will be beneficial to a powerful group of interests they may only change by social upheaval.[3]

Analysis of the relation between change in institutions and market development has particular relevance for the study of change in the role of the state in agricultural markets. In the mid-twentieth century the state intervened in agricultural markets well beyond the extent that market failure required (see Chapter 2), setting up restrictive 'rules of the game' – which became increasingly out of line with wider market realities as markets developed, at the cost of economic growth. The subsequent liberalisation is only being accomplished through collective social action and involves much upheaval. Where liberalisation has occurred, prices of resources and products have been brought more into line with world prices, as a result of trade and investment flows. A challenge now is how to adjust governance of agricultural markets in countries where institutional change is constrained politically and bureaucratically, and where liberalised agricultural markets demand responsiveness from government for benefits to be realised.

Institutional analysis has deepened in recent decades, developing from several streams of enquiry (firms, markets, information, politics, history) which together loosely make up the 'new institutional economics' (Hubbard 1997a). Institutional economics[4] traditionally reacted to mainstream economic theory of the consumer and the firm, arguing that the motivation of consumers and firms was more complex than the utility and profits maximisation portrayed by mainstream theory. The 'new institutional economics' is intent on analysing the two-way relation between institutions (the conventions, authority, laws and regulations which make up the 'rules of the game' in society at any time) and markets (buying, selling and assuring). Organisations (firms, bureaucracies) facilitate production and exchange and mediate the relation between markets and the institutional environment – and contain that relation within them, through the agreements for exchange and use of human and material resources among those in the organisation. Thus over time organisations influence markets and institutions, by changing how they do things (technological change) but at any one time they are governed by the prevailing 'rules of the game' of the society in which they operate.[5]

Much of the study focuses on internal adjustment within state organisations, in response to liberalisation moves. Here tools (such as principal-agent theory, transactions costs, moral hazard) which have become part of NIE's armoury for analysing organisations prove useful in the field investigations.

State failure

State failure is said to occur where state intervention causes greater cost to society than the market failure which it was designed to correct.

Where the state is efficient, market failure provides a case for state action, so that the supply of goods and services can be brought into line with what people want. But where the state imposes greater costs than the market failure it is trying to reduce there is no case for collective action. Better live with the market failure and hope for market development (i.e. improvement in the functioning of markets) to occur, or wait for political change to a more market friendly regime and more efficient collective action.

The theory of government failure (Wolf 1988) argues that the state is inherently less efficient than private production for four reasons. First, there is a disjunction between costs and revenues. 'Where revenues that sustain an activity (such as taxes) are totally unrelated to costs or output quantities, the probability that costs will not be controlled and that output levels will be unwarranted is much increased' (FAO 1991: 112). A lack within the state of an automatic mechanism to contain costs within revenues causes the disjunction. Resource allocation in the public sector is 'essentially a political process characterised by lags, bottlenecks, coalitions, log-rolling and other fuzzy attributes of political behaviour' (Wolf 1988: 62, quoted by Walsh 1995: 24). Second, the public sector lacks objective measures of performance. Third, there are external costs of public interventions – often taking the form of increased transaction costs in doing business (e.g. the costs of obtaining licences and permissions). Finally, government interventions tend to serve the material interests of those in powerful positions in bureaucracies rather than reducing the distributional inequities for which they were intended. This echoes public choice theory (e.g. Buchanan 1986) which argues that bureaucracies are essentially self-serving rather than dictated by the ethic of service to the public, as in Weber's ideal bureaucracy.

The debate over 'state failure' theory focuses on the extent to which the potential failings of the government identified by Wolf are inherent in the operation of the public sector.[6]

The 'new public management'

Emerging from the concerns expressed in the state failure debate are recent programmes of public management reform which attempt to bring market mechanisms into public decision-making – referred to as the 'new public management' (NPM) approach. NPM has been the most influential guide for public management reform programmes in recent years. Progress towards realising the NPM model has varied greatly, with the most committed implementation in the countries which were in the vanguard of radical privatisation, such as Chile, New Zealand and the UK.

Underlying NPM is the ideal of the 'enabling state', which carries out only those tasks which the private sector cannot, supports market development, and balances its budget (in an average year) so that currency stability is underpinned and private savings are available for private investment. In the NPM model the state confines itself to policy, planning and financing services. Competition among service providers bidding for state contracts to deliver public services is the basis for raising performance.

Contracts are central to NPM. Contracting as an approach to the management of organisations became popular in the private sector in the 1980s as a means of increasing the efficiency of large firms, by enabling them to concentrate on their core functions, and buying in all other material inputs and services (Walsh 1995). Applied to the state the implication is that government moves towards a structure in which semi-private executive agencies of government surround a much reduced civil service and local government core; the activities of the core being to render account to the public, prepare policy, and supervise contracts with the satellite agencies and the private sector, which deliver the public services. In short, the implementation agencies move out of the civil service on to contracts, and compete alongside private firms for government work.[7] Competition is at the centre of NPM as a philosophy and strategy. The idea is that incentives are increased for public officials and agencies to act efficiently and accountably by setting and enforcing clear contractual goals and penalties.

As a model for reform of public administration, NPM both accords and conflicts with the previously dominant ideal – Weber's public-spirited bureaucracy. For the core public functions NPM requires an efficient bureaucracy, hierarchically accountable to political decision-makers. But for all non-core functions accountability is via contracts, with the stress on reducing the costs of outputs by using competition for contracts among providers.

In sum, the strands of theory reviewed above suggest public goods (i.e. those which can only be provided collectively) vary by circumstance and culture, with the exception of a few pure public goods. Administrative arrangements for provision of public goods reflect the institutions ('rules of the game') of the particular country. These arrangements are likely to change only slowly.

The state failure debate and public choice theory allege an inherent tendency to inefficiency and self-serving in public service bureaucracies. NPM, now the dominant model for public services reform, attempts to raise efficiency by bringing market mechanisms into public services. Previously, public services reform focused on raising quality of management and personnel, on the assumption that private firms would not provide the service or would exploit consumers.

The theoretical observations above lead directly to the main study question of the research: whether in the institutional circumstances of particular countries, NPM provides the best means for reforming public service delivery.

We now move to a more detailed discussion of tasks for government in implementing NPM. The focus is on commercialisation of public services, since NPM is at the heart of the broader move to commercialise public services in order to increase the power of consumers, and to use competition to increase quality and reduce costs.

Commercialising public services: tasks for government

Governments in many countries engaged in market liberalisation and structural adjustment are changing their organisation of services provision to make services more consumer responsive, less costly to the government's budget, delivered to an increasing extent by independent state-owned agencies, and by private and voluntary organisations. Contracts between government and service providers are an increasing feature.

Market liberalisation and structural adjustment are associated with these changes through the emphasis they give to consumer choice, and the resulting increase in the number and capability of private providers (particularly in processing, trade and services) and in competition among them.

Choices in service provision structures

While the desire to achieve greater market and consumer orientation in public services provision is becoming universal, the chosen methods to achieve it differ. The main choices are:

- centralised vs. decentralised service provision structures
- direct provision within civil service structures vs. indirect provision by contracted private and non-government organizations.

Existing structures for provision of public services in many adjusting countries are centralised, and services (e.g. research experiments, extension advice) are delivered directly by government employees. Therefore the choice of future structures is determined by the capacity for achieving greater consumer orientation and value for money in existing centralised structures, versus the potential for good management of decentralised decision-making structures, contract making, monitoring and enforcement. Figure 1.1 illustrates the choice.

Other factors in the choice are the speed with which commercialisation and cost reduction need to be achieved and the overall capacity of the civil service.

The public agricultural services of many countries have been built up on a centralised, regulated, non-commercial basis. Chains of command are often long. Financial management is centralised. These characteristics have made performance vulnerable to resource losses (staff and operating budgets) during the financial stringency of structural adjustment. In many developing countries, public agricultural service structures have remained substantially unchanged despite considerable development of:

- input and product markets resulting from market liberalisation
- alternative sources of agricultural advisory support
- increasing non-agricultural income reliance of rural people.

<table>
<tr><td>

Centralisation vs. Decentralisation

1 Capacity to manage decentralised decision-making structures

2 Capacity to achieve greater consumer orientation and value for money within existing centralised structures

Direct vs. Indirect Provision of Services

3 Capacity to manage indirect provision well (making, monitoring and enforcement of outsource contracts)

4 Capacity to achieve increased consumer orientation and value for money with direct (i.e. in-house) provision of services

</td></tr>
</table>

Figure 1.1   Capacity factors in the choice between centralised and decentralised service management structures, and between direct and indirect provision

Consumer oriented reform of public agencies

Consumer oriented reform of public agencies is a four-part process, in which government:

- gives more power to the consumers of public services (consumer attitude surveys, defined service standards, consumer complaints lines, consumer associations closely involved in its decision-making)
- allows public service delivery agencies to develop their commercial potential and enjoy increasing managerial and financial independence, and ultimately obliges them to compete with external agencies for contracts to deliver public services
- removes market unfriendly regulations (i.e. any regulations which prop up private or public monopolies) and promotes market friendly regulation (e.g. food safety, quality assurance) to be carried out as far as possible by stakeholders in the industry (consumers, traders, processors, producers) in a board or council enforcing acceptable standards, funded by a self-imposed levy
- strengthens its own three core public functions which facilitate service delivery (policy development, services management, regulatory overview).

Corporatisation, privatisation, unbundling and contracts

Corporatisation is the term given to changing a government department into an autonomous public enterprise, subject to commercial law on taxes, accounting practices, competition rules and employment. Employees are transferred from civil service status to commercial contracts.

Restructuring of a public agency may be required for it to be corporatised or privatised. This involves unbundling its activities in order to separate those which can be run privately and competitively from those which are more strongly public in nature (e.g. natural monopolies) which may be retained in the public sector or closely regulated. Vertical unbundling, or vertical disintegration, has been widely used as a means of breaking up vertically integrated public utilities (e.g. splitting electricity generation from electricity distribution). Horizontal unbundling splits a service geographically (e.g. into different electricity distribution zones) or by service category (e.g. separating passenger and freight services in the case of railways).

Explicit contracts between the corporatised service provider and government have been used 'to increase the accountability of employees and managers and to improve the focus of operations by clarifying performance expectations and the roles, responsibilities and rewards of all those involved' (World Bank 1994b: 42). For example, performance agreements specify clear targets (e.g. productivity, service quality, administrative and financial), obligations of government, means of monitoring performance, and provide clear incentives and sanctions (e.g. bonuses, fines). They are widely used as the basis for regulating a public service carried out by a private or public provider.

Sequencing of cost recovery and commercialisation

Reform requires setting priorities for change, often referred to as sequencing. An example of sequencing for commercialisation of public services is shown in Figure 1.2, drawn from discussions for reform of agricultural services in Zimbabwe.

Where a department performs an essential function for an industry (e.g. quality control) collaboration with industry associations may provide a way to commercialise and reduce public costs. The preparation for commercial operation of Dairy Services, a department in the Zimbabwe Ministry of Lands and Agriculture, is an example which illustrates this approach (see Chapter 11 below).

STEP 1	For all service-providing departments	Revenue retention and easy access to funds enabled. 'Revolving funds' may be used for this purpose. Accounts reoriented to cost centre management (including costs of staff, fixed assets and operation). Business plan in action, setting targets for cost recovery and service standards to be achieved.
STEP 2	For service deliverers with good revenue potential	As cost recovery and commercialisation increase, move towards a contract arrangement, in which retention of the contract is dependent upon maintaining/improving standards of service delivery, in exchange for budgetary funds.
STEP 3	For service deliverers able to manage independently	Service delivery organisation is made fully commercial, ends civil service status, and competes for government and other contracts.

Source: Hubbard (1997b)

Figure 1.2   Example of an implementation sequence in cost recovery and commercialisation of public services

Challenges in increasing cost recovery

Finance: Enabling public agencies to retain a portion of the revenue they collect is a key step in encouraging cost recovery. The means used by the Ministry of Finance in Zimbabwe in the mid-1990s to encourage cost recovery was to set up extra-budgetary revenue retention funds (revolving funds). Imprest accounts[8] were also introduced to facilitate payment of expenses to staff. Strengthening financial management capacity is required to give any attempts to commercialise an agency's services a chance of success.

Staff incentives: Commercialisation often requires staff to work harder to a higher standard, and to take more initiative. Besides training staff to market-oriented approaches, material incentives (bonuses, pay increases, promotions) beyond those possible within the existing performance appraisal arrangements are usually essential.

User charges: User charging is essential for commercialisation. But it is controversial in the case of charging low income users. Charges need not be full cost: it is possible for government to contract with the service provider to offer services to some people or areas at below cost, with government paying the difference – for example, the dipping fees which Zimbabwe's Veterinary Services Department began to charge communal farmers in the mid-1990s.

Contract management capacity: The process of commercialisation of services provision involves the expansion of contracting as the basis for service delivery. Contract management involves tender design, transparent tendering processes, and monitoring of contract implementation and performance. Without developed contract management processes delays and corruption reduce the benefits of service commercialisation. Building contract management capacity will be important in both service provider agencies and in government departments which are using the contractors.

Estimating performance in cost recovery: Commercialisation requires substantial managerial independence and knowledge of the full costs of activities. Cost recovery is broader than user charging alone, being earnings on all assets allocated to provision of a service.

Capacity requirements to commercialise services successfully

Having overviewed the tasks for government in commercialising public services, with examples from agricultural markets, we now turn to the capabilities (capacity) required to undertake such tasks. The study focused on the capacity of the state to pursue a market-enabling policy agenda, to provide core public services, to commercialise services and to manage the commercialised services effectively.

Three broad categories of state capacity were identified for the purposes of the present study:

1 *Political commitment*: This refers to the commitment and power of national leadership to pursue a market-enabling policy agenda.
2 *Policy development*: Ongoing formulation and revision of policy in order to attain objectives.
3 *Effective operation of public agencies*: Ability to implement well and adapt. Management criteria used included leadership, organisational autonomy, management and administration, commercial orientation, consumer orientation, legislative framework and organisation culture.

The approach to capacity taken in the study is broad, linking analysis of organisations to their institutional context. It covers:

- the circumstances of the introduction of reforms
- the appropriateness of reforms to the level of development of state and market
- the presence of the skills, administrative structures, forms of management and levels of plant/equipment needed to fulfil the intended organisational (contractual) arrangements
- the availability of information and human resource management to create incentives for efficient and responsive behaviour
- the institutional conditions needed to support the form of organisation/contract adopted and to release its potential for efficient and responsive behaviour (Batley 1997: 20).

Approach of this study

Liberalising agricultural trade changes the public services needed for successful development of agricultural trade. Some existing services (e.g. public marketing agencies) may have little remaining public role, while there may be an expanded need for other public services (e.g. meat hygiene). There is a need for adjustment in the public services provided.

The purpose of this study is to throw light on why and how governments in developing countries adjust or fail to adjust public services to agricultural trade after liberalisation of trade. In examining how governments adjust their public services to agricultural trade, the focus is on whether the 'new public management' provides the appropriate model for public services post-liberalisation of trade, and the appropriate methods for reform.

This chapter discussed the 'new public management' and commercialisation as a method of reforming public agencies. It also introduced concepts used regarding private and public goods and institutions. The changing role of government in agricultural markets worldwide is discussed next. This is followed by analysis of the changing role of government in agricultural markets of each case-study country. Attention is then turned to particular issues in reform of public services to agricultural trade: food security, commercialisation of public marketing agencies, ensuring quality in marketed produce, and the public role in information for agricultural producers, traders and consumers. On each issue examples are drawn from the case-study countries. In the light of earlier chapters, the final chapter discusses factors determining whether governments have been able to adjust their public services successfully after liberalisation of trade, and the appropriateness of the 'new public management' model.

2
Reforming the Role of Government in Agricultural Markets

For most of the twentieth century, particularly after the 1930s depression, governments worldwide intervened substantially in both input and output markets for agriculture, in some countries excluding private firms altogether. Seed, feed, fertiliser, pesticides, finance and advice were provided by state organisations or with state subsidies. Fixed asset investment (e.g. land purchase and development – dams, roads, fences) was often similarly assisted. Crop purchase was often carried out by widely distributed state depots offering standard prices across a whole country (pan-territorial pricing). Grain stocks were accumulated by the state in order to set and stabilise consumer prices. Large bureaucracies were set up to administer public agricultural services, imposing much budgetary cost – particularly for price stabilisation, mostly grains. The main purpose was to achieve self-sufficiency in food staples or to promote exports of agricultural commodities.

In several states, e.g. in the European Union, India and China, substantial state intervention persists, particularly in state buying and selling and import restrictions. But in all cases there are moves to reduce intervention by the state in markets for agricultural outputs and inputs, as well as more widely in the economy. At the same time there is a worldwide shift of emphasis by states towards consumer welfare – motivated above all, in agricultural markets, by consumer concerns regarding food safety.

This chapter[1] discusses why states have intervened so extensively in agriculture, why reform of public agricultural services tends to be slower than liberalisation of agricultural trade, and how liberalisation is changing the intervention needed. It provides background for the discussion in later chapters of particular issues regarding adjustment of the public role in agriculture after liberalisation.

Why governments intervened extensively in agricultural markets

Agricultural economics texts in the mid and late twentieth century tended to emphasise that the underlying reason for extensive state intervention in agricultural markets is the high degree of market failure[2] from which they suffer: agriculture is regarded as being particularly prone to market failure because food is an essential,[3] and production is liable to disruption by disease and weather. Together, these combine to create unstable prices. Transaction costs tend to be high where there are many small producers – as in farming – requiring common, if not collective services, to reduce them. The external costs (i.e. costs to society) of high-cost, unstable agriculture and food prices are argued to be unacceptable.

But the determination from the mid-twentieth century to protect and develop domestic agriculture requires more than a technical explanation for when and why it occurred. The fashion for nationalism (discouraging trade, emphasising national self-sufficiency in food) and for state control, need to be part of any explanation. State intervention in agriculture arguably went well beyond what was needed in order to prevent failure of agricultural markets.

Purposes of intervention

Four general considerations motivated the extensive state intervention in agriculture in the twentieth century:

1 *Food self-sufficiency.* While the focus of food policies has shifted in recent years to a concern with household food access, for many years food security was interpreted as synonymous with security of national supplies. In consequence, many countries implemented policies designed to achieve domestic self-sufficiency in food crop production. In addition, the establishment of public reserves was regarded as necessary for a prompt response to food emergencies.
2 *Stabilising prices.* In the belief that price variability deters investment by risk-averse producers, government efforts to enhance the adoption of new technologies have centred on the offer of guaranteed producer prices.
3 *Promoting welfare.* In many countries a policy of subsidised consumer prices has been used, with the main beneficiaries tending to be

politically volatile constituencies, such as the urban population, civil servants and the military.

4 *Substituting for private intermediaries.* A belief that the private sector is poorly developed and therefore unable to perform well has also prompted government intervention.[4] In addition, deeply held views about the exploitative motives of traders and popular resentment of non-indigenous communities dominating distribution channels has been an important force.[5]

While these four factors shaped policy, subsidiary objectives have been relevant, especially the extraction of surpluses from the agricultural sector in order to channel funds into other areas, notably the manufacturing and public service sectors.

Of the four motives outlined above only the second and the last can be interpreted as intervention for the promotion of marketing efficiency (i.e. to reduce market failure). The other two relate more closely to social welfare and redistributive objectives.

Controllers and implementers of intervention: the agricultural ministry and state agencies

Control over interventions in agricultural markets was usually put in the hands of an agricultural ministry with state agencies as implementers. Where the state took over substantially the supply of inputs and distribution of outputs, state agencies dominated markets – for seeds, fertilisers, agricultural finance and particularly for distributing agricultural production domestically and externally. Dominance of state agencies in agricultural markets was greatest in two types of country: first, countries reliant on commodity exports and those where large-scale commercial farming was politically powerful, often with a strong motive to protect domestic grain markets from external competition. Western Europe, Australia, New Zealand, South Africa and Kenya are prominent examples. State agencies dominant in these agricultural economies were substantially producer-controlled and often linked closely with cooperatives dominated by large producers. Second, in communist countries, where state production and distribution substituted virtually entirely for private activity. State agencies in communist countries were dominated by state bureaucracies, which also ran state farms.

At the centre, nominally controlling the state agencies and responsible for agricultural policy, has been the agricultural ministry – dedicated to the welfare of farmers, often the largest farmers. Agricultural policy focused on decisions regarding the extent of direct intervention.[6]

Services to agriculture (e.g. agricultural extension, veterinary services, research and laboratories) were sometimes taken on by state agencies – usually for particular commodities – but more usually remained as departments within agricultural ministries. As consumer rather than producer needs now become dominant, the very existence of agricultural ministries is threatened – some countries have already replaced them with rural development ministries.

Intervention instruments – controls on trade, state trading and stockpiling

In order to attain some or all of the four objectives above (self-sufficiency in staples, stable prices, welfare and replacement of private trade), or subsidiary political objectives, states have intervened in the trade in agricultural outputs and inputs. The stabilisation of consumer and producer prices has been a dominant objective of government intervention in many countries, commonly implemented through state agencies. Domestic populations have been a dominant insulated from price variability on world markets with state agencies retaining the monopoly on external food trade. In addition, domestic prices have been fixed for both producers and consumers. The main instruments used are controls on trade, state trading and stockpiling of staples.

Controls on trade

Governments in many countries have placed a high priority on being able to exercise control in agricultural markets. Price control has been an important objective, particularly for staple food. In some countries controls reflect their vulnerability to political upheaval in the event of food price inflation; in others the desire to favour particular interest groups. Where the concern is price stabilisation, controls have taken the form of licences, restrictions on external and inter-district movements, as well as quantity and storage controls. Together these measures were believed to facilitate not only the operation of government pricing policy, but, more importantly, government's ability to command food supplies in periods of shortage. Many governments have retained the right to confiscate private stocks in the event of a food emergency.

State trading

State agencies were charged with the wholesale function of assembling and transporting agricultural products or inputs via widely dispersed depots, often at uniform prices across the country. The prices were set to achieve policy objectives rather than to enable the corporation to cover its costs. State agencies would either participate in the markets concerned, trading alongside private buyers and sellers, or operate a 'single channel' state trading monopoly.

Stockpiling

While trading requires some inventories of the product traded, stockpiling (particularly of grains, but also of other products in some countries – famously beef and wine in the case of the European Union) is designed to fulfil policy objectives. The maintenance of grain reserves has been a feature of food market interventions in many countries. These stocks are used to stabilise consumption in the event of shortfalls in food availability, to maintain domestic producer prices above market level for some or all seasons, and to stabilise domestic consumer prices of grains – particularly for urban dwellers, public officials and armies. While it is possible to use imports to stabilise domestic consumer prices, many governments have been reluctant to rely on international markets because of fears about their unreliability and the foreign exchange costs which would be incurred. The widespread accumulation of substantial strategic grain reserves has required large budgetary outlays for buying, storing and distributing stocks.

Regional variations in intervention and reform

While public intervention in agricultural markets has occurred worldwide, there is much variation among countries and regions in the extent of intervention. The case of grain marketing interventions in developing countries is indicative, as illustrated by Table 2.1. The importance of such variations for this study is that they determine in part the challenges and process of reform of the state's role in agricultural markets following liberalisation of agricultural trade. The pattern of pre-reform experience inevitably determines post-reform possibilities.

In Table 2.1 pre-liberalisation intervention in grain markets by a selection of countries from Asia, Africa and Latin America are compared. The existence of a public marketing agency is common across regions but the tasks of those agencies and the environment in which they operate have varied widely. At one extreme are boards with statutory monopolies over domestic and international marketing (e.g. Zimbabwe) and at the other are boards which not only coexist with private trade but also provide services, such as the publication of market price data, which encourage private participation (e.g. Sri Lanka).

Table 2.1 indicates that the existence of a public marketing agency transcends ideology, economic system and historical experience. It also

Table 2.1   Cross-regional comparisons of grain marketing interventions pre-liberalisation

	Public marketing agency	*Statutory monopoly*	*Internal movement controls*	*Price support/ price ceilings*	*Buffer stocks*
Asia					
China	Yes	Yes	Yes	Yes	No
India	Yes	No	No	Yes	Yes
Indonesia	Yes	No	No	Yes	Yes
Korea	Yes	No	No	Yes	Yes
Sri Lanka	Yes	No	No	Yes	Yes
Africa					
Ghana	Yes	No	No	No	No
Mali	Yes	Yes	Yes	Yes	Yes
Tanzania	Yes	Yes	Yes	Yes	Yes
Zambia	Yes	Yes	Yes	Yes	Yes
Zimbabwe	Yes	Yes	Yes	Yes	Yes
Central America					
Costa Rica	Yes	No	No	Yes	Yes
Nicaragua	Yes	Yes	Yes	Yes	Yes

Sources: There is an extensive literature documenting country-specific policies. These include: for China (Findlay *et al.* 1993); for India (Radhakrishna 1988); for Indonesia (Ellis 1993); for Korea (Bai-Yung Sung 1993); for Sri Lanka (Gunwardana and Quilkey 1988); for Ghana (Coulter 1993b); for Mali (Thompson and Smith 1991; Coulter 1993a); for Tanzania (Nindi 1990); for Zambia (Government of Zambia 1993); for Zimbabwe (Jones and Beynon 1992); for Costa Rica (Hazell and Stewart 1993); for Nicaragua (Spoor 1994).

suggests differences among regions in the extent to which the agency has been expected to substitute for the private sector. In many African countries public marketing agencies had a statutory monopoly on food trade, a position often backed by restrictions on internal movements of food. In contrast, state marketing agencies in other regions have largely been expected to supplement the private sector, at least in internal markets. In all regions pricing policy has been a major motive for intervention with most countries using both producer price supports and consumer price subsidies. There are of course divergences of countries from these broad regional generalisations.

Differences in pre-liberalisation arrangements make for differences in the extent of adjustment in the state's role needed after liberalisation of trade. Where marketing interventions were designed and operated in opposition to the private sector greater adjustment is needed by the state if it is to take on a market enabling role successfully. Arguably this has been the situation in much of Africa. Successful adjustment in the role of the state in such cases requires revision of entrenched views about the motives and capacity of the private sector.

Why states are liberalising agricultural markets

Since the late 1980s there has been great pressure on states to liberalise both their domestic and external agricultural trade. External pressure has come mainly from the World Bank and International Monetary Fund arguing that control of markets by the state is distortionary and generates significant economic inefficiencies. External pressure has been exerted through conditionality attached to financial assistance.

The impetus for reform from within countries themselves came from various developments. In some former communist and Eastern bloc countries it was part of the wider shift in policy towards a greater role for resource allocation by private markets. In others, reform was forced on governments by the failure of existing state organisations to successfully implement pricing policy and marketing services. For example in Tanzania, Zambia and Zimbabwe reform was driven by the financial crises in the state grain marketing agencies. In the countries of the former Soviet Union there was a wholesale collapse of the pre-existing system. In all countries there was dissatisfaction with the mounting budgetary cost of supporting loss-making parastatals, and frequent allegations of corruption in state marketing agencies (e.g. in Kenya and India). Therefore the external pressure by donors often found support from opposition groups within countries.

The financial burden of existing interventions has been crucial in paving the way for reform, either because of the pressure from domestic political constituencies or because reliance on external aid has left many countries open to substantial leverage by donors. In this context it is important to note that where countries have not faced recurrent financial crises they have generally not embarked upon substantial programmes of reform, particularly in grain marketing. India, Malaysia, South Korea and Indonesia before the financial crisis of 1997–8 all fall into this category. The often more rapid but incomplete reforms in some African countries result from lack of resources to sustain existing systems, donor pressure for reform, but absence of a sufficiently powerful political coalition to sustain reform (e.g. Kenya, Zimbabwe, as discussed in the case studies below).

In sum, it is arguable that the impact of mounting budgetary cost of price policies to subsidise agriculture has been inversely proportional to the wealth of the state: the poorer the state the greater the pressure from budgetary costs to force through change. Thus one of the world's richest regions (the European Union) has been able to afford to continue substantial protection of its agriculture, mainly through trade interventions (the variable tariff, a variety of quantitative restrictions on imports, plus subsidisation of exports) despite the high budgetary costs. Only since the formation of WTO in 1996 has change in European policy become unavoidable, because of the timetable for enforcement by WTO of members' commitments to liberalise their agricultural trade.

In poor countries with a high level of intervention – many of them in Africa – trade liberalisation was deepest, owing both to spiralling budgetary costs of their food policies and to their high level of dependence on international assistance through structural adjustment programmes, which insisted upon liberalisation of domestic and external trade as a condition of assistance. By 1994 all twenty-eight countries in Africa undertaking adjustment (as classified by the World Bank) had removed the major restrictions on market participation (Jones 1994). Prior to this fifteen of these countries had significant restrictions on private trade and in a further five countries public marketing agencies were important players in crop markets. While the fiscal drain of state marketing agencies is widely acknowledged in the case of Sub-Saharan Africa, they were no less a burden in other developing regions. For example, in the mid-1980s the share of the annual government budget devoted to handling grains was no less than 3.5 per cent in Mexico, 4.6 per cent in India, and 10.5 per cent in China[7] (Knudsen *et al.* 1988: 53). Nor have countries in other geographical regions been immune from the pressures exerted by the international financial institutions; the experience of the Philippines, Nicaragua and Bangladesh are examples. In short, for those countries heavily reliant on foreign aid flows, the option of continuing to finance large-scale public interventions in marketing has no longer been feasible.

Whether the reform process has been 'pulled' by internal factors or 'pushed' from the outside, two influences have been dominant: namely, the unsustainable financial burden of existing interventions, and the failure of these to achieve their intended objectives.[8]

The unsustainable financial burden

Why did the price stabilisation arrangements set up in so many countries from the mid-twentieth century usually lead to growing losses and substantial financial burdens on the public purse, particularly in the case of grain staples? Food price policies of the state were the main cause, together with organisational arrangements which encouraged waste and corruption:

- There was a tendency to stabilise producer prices for grain staples above market prices, leading to overproduction and cultivation of grains in areas less suitable for them, especially since alternative crops were often underpriced – particularly fibres for local industry, e.g. cotton. Where pan-territorial grain prices to producers were applied, overpricing of grains was worst in remoter areas.
- A tendency to maintain consumer prices for locally produced grain staples below import prices forced trading losses on state marketing agencies.
- Financial losses were inevitable on non-commercial functions (mainly emergency food distribution) carried out by state grain marketing agencies with no reimbursement.
- Pan-seasonal prices for producers put the grain storage function on to the public organisations. Uniform prices throughout the year discourage private sector investment in storage capacity because there is no return to capital tied up in stocks. This leaves storage to become effectively a public sector activity. In addition pan-seasonal pricing creates severe pressure on transportation, distribution and storage facilities because it causes farmers to concentrate sales in the immediate post-harvest period.
- Buying through widely dispersed depots put much of the crop transport costs on to the state marketing agencies.
- Waste, excess staff and theft in state marketing agencies have been major problems.
- The soft budget constraints under which state agencies operated, with government underwriting not only trading losses but also commercial loans, further reduced incentive to improve performance.
- Being part of the public sector, the state agencies were under pressure to provide employment opportunities, thus leading to overstaffing and high wage bills.

Failure to achieve objectives

In addition to the fiscal cost of interventions, pressure for reform has arisen because of the failure of these interventions to achieve their objectives. There have been few examples of effective price control in those countries where the public marketing agency has enjoyed a statutory monopoly. The parallel, and often illegal, markets handled a large proportion of marketed output. In many countries 'liberalisation' has meant the government officially recognising the *de facto* participation of the private sector. Prices in these channels have been inflated by the need of traders to dodge or bribe their way through official controls.

The poor management of the public marketing agencies, plus their precarious financial position (they often lacked liquidity at harvest time) meant that many producers simply had no access to marketing services. Even where they did, the tendency towards late or inadequate payment by the public marketing agencies caused serious difficulties. Consequently, the coverage by public marketing agencies was often patchy and unpredictable.

In sum, states are liberalising their agricultural markets because liberalisation offers the potential to relieve them of substantial budgetary costs, to create more effective services for farmers and consumers, and to remove the obstructive and rent seeking power of petty officials.

By the mid-1990s state monopolies of external and internal trade in agricultural outputs and inputs had been largely removed. Internal barriers were the first to go, followed by a more gradual and partial lifting of external barriers (e.g. the Kenyan removal of NCPB's maize import monopoly). The fears of price instability were allayed, once import trading arrangements had settled down,[9] as in Kenya (Lewa and Hubbard 1995). A new *modus operandi* for agricultural trade was being established in those countries where it had been most restricted. The dismantling of both trade restrictions and state marketing boards in South Africa is the most influential case (Bayley 2000), with favourable results to date in trade, storage and price stability as the private sector has taken up a larger role.

A general consequence of greater import liberalisation is that imports now have a larger share of the market in many countries, particularly in major coastal cities and regions, often keeping staple prices lower than the levels at which they had previously been maintained. The cost of adjustment as the state marketing agencies reduce their operations has been felt in remoter rural areas where grain sales points have been lost to farmers and not replaced by private trade (or only at much lower prices). Zimbabwe, Kenya and Sri Lanka are cases in point, where farmers – particularly in remoter areas – have faced lower producer prices for grain as a result of the grain marketing agencies cutting their unprofitable buying operations, and as a result of imports. This has resulted in much political agitation by farmers for protection from imports, incomplete reform of Kenya's NCPB, and imposition of variable import tariffs on maize imports to Kenya and rice imports to Sri Lanka.

Why reform of public agricultural services tends to be slower than liberalisation of agricultural trade

In general, the reform of public agricultural services has been much slower than liberalisation of agricultural markets. A few high income countries in which the 'new public management' has been particularly influential on policy – New Zealand, United Kingdom, Australia – quickly scrapped their state marketing agencies and privatised or contracted out much of their public agricultural services (e.g. extension, laboratories, veterinary services). But in many developing and transition countries the state apparatus has been reformed only slowly, despite its increasing inappropriateness to liberalised markets for agricultural inputs, products and services. Examples include Uganda in the 1990s (Belshaw, Lawrence and Hubbard 1999) and Zimbabwe and India as discussed in this volume.

Broad reasons for slow change in the state apparatus for agricultural services include:

- Continuing lack of confidence in reform policy. This makes retention of a degree of control over markets – particularly for staple foods – a continuing political imperative for some governments.
- State grain marketing agencies had the dual function of stabilising prices and acting as the state's emergency arm for food distribution to distressed areas during emergencies (e.g. droughts, floods). Some governments want to retain grain marketing agencies for emergency purposes.
- The organisations and institutions created to administer public agricultural services are often politically difficult to reform, because of vested political and bureaucratic interests in them and sometimes rents which existing arrangements allow individuals to extract.
- The capacity and willingness of the private sector to assume the marketing functions being shed by the state is unevenly distributed along the marketing chain with the weakest link often in the storage function and in serving remoter farming areas. This keeps alive a lobby for preserving state agencies carrying out these functions, even when they are no longer able to do so.

How liberalisation is changing the interventions needed in agricultural markets

While liberalisation of agricultural markets is relieving states of the financial burden of previous policies, it creates new hazards (e.g. private monopolies replacing public, risk of greater food price instability in times of crisis) and new roles for the state to perform (in helping to develop agricultural and food markets and ensure that they function well, particularly in remote areas, and achieving food security without restricting trade).

Market liberalisation requires changes in market governance. Specifically, when production and trade controls are removed or greatly reduced (i.e. liberalisation) changes in government activities are required in order for society to benefit from liberalisation (i.e. to receive better and cheaper goods and services). These changes consist in:

1 taking on new public tasks to promote market development (infrastructure, facilitation and regulation of markets to promote competition)
2 reducing provision of services by government which no longer need to be financed by public expenditure. These are services:

   (a) which can be provided better and cheaper by private firms in competition with each other, and
   (b) which, through being subsidised or supplied on a state monopoly basis, hinder ('crowd out') the development of private sector services.

Therefore governments are faced with the challenge of entering new policy areas while reducing their involvement in some traditional policy areas. Table 2.2 below summarises these policy implications.

It is noteworthy that to facilitate market development the policy-making capacity of government is key – reviewing the impact of past policy, considering options and (above all) bringing stakeholders into active consultation on policy development. During the era of controlled markets the policy development function in agricultural ministries was internally focused on government's control of prices and buying and selling of key crops and inputs. In a liberalised market environment, policy needs to be focused on market development, seeking to deepen and widen markets, in order to encourage investment, quality assurance and integration of dual agricultures (small and large farms, commercial and subsistence).

Conclusion

This chapter discussed why states have intervened so extensively in agriculture, why they are liberalising markets, and how liberalisation is changing the interventions needed.

Table 2.2   Impact of liberalisation on role of government in agricultural markets

Policy areas	*Type of policy*
New agricultural policy areas to be entered:	
1 Encouraging investment in agricultural production and trade by promoting confidence in markets	• Promotional (stability, predictability, transparency) • Assurance (quality)
2 Maintain competition in markets	• Deregulation of external trade • Regulation of monopolies
3 Integrate agricultural sector if dualised (e.g.commercial and communal farming)	• Infrastructure, extension, research, rural development
Old agricultural policy areas in which to reduce involvement:	
4 Public services in competition with private sector	• Internal reorganisation of agriculture ministry • Commercialisation and privatisation
5 Emergency relief assistance (e.g. in droughts)	• Income based redistribution (e.g. public works) replaces food and seed distribution

In sum, the reform objective for agricultural markets now tends to be similar for all countries – to build a vibrant, competitive private agricultural marketing sector. The issue is rather how to achieve it. There is a range of options for providing public services, and for how government can take up new tasks and phase itself out of others where its providing role is no longer needed. There are further options in the policy process – whether to make decisions on a unilateral basis or together with associations of stakeholders in the sector. But in practice the options are constrained. The greatest immediate constraint is political support for change, since there is inevitably resistance from those benefiting under existing arrangements, such as farmers receiving subsidies, traders with preferential market access, and state employees who fear redundancy. An underlying constraint is often the state's ability to reform previous policy, manage its own resources – its personnel and its finances – and to implement its own new policy.

In the following chapters, the response of governments in the case-study countries to the challenge of liberalisation of agricultural markets is the focus. The stress of the analysis is on how states adjust their services in response to liberalisation of agricultural markets. The response by states is determined by the style and extent of their previous intervention, as well as by the overall capacity of the state to adjust its priorities and bureaucracies. Where exclusion of the private sector was greatest (former communist countries and some African countries) there has been more adjustment needed. The greatest challenge is faced by states with a low capacity to adjust and with poorly developed markets for agricultural inputs and outputs.

Part II

Country Studies of the Changing Role of Government in Agricultural Trade

3
India

Renu Kohli and Marisol Smith[1]

Introduction

The total population of India is approaching one billion, making it the second most populous country in the world. There is enormous ethnic diversity and about a third of the population is classified as poor. India achieved independence in 1947. Its constitution provides for party government, with a legislative assembly, an upper house, and a permanent civil service. The country is a federal collection of twenty-five states and seven union territories, with local government as the third stratum of government. The constitution permits the centre to impose its will on the states by declaring a constitutional emergency.

After independence India pursued a policy of self-reliance, following the dominant Congress Party's socialist vision of development. India developed along mixed economy lines with a large public sector, and a private sector under a system of licences and direct controls. The private sector was shielded from competition and the domestic economy insulated from the world market. This pattern dominated the post-independence period until reforms introduced in 1991.

In early 1991 India faced a macroeconomic crisis. The fiscal deficit was 10 per cent of GDP, the current account deficit was 3 per cent, inflation was 12 per cent and rising, and the decline in foreign exchange threatened to leave government unable to service its foreign borrowing. Assistance from the IMF was sought. In July 1991, the traditional model of Indian development in which a dominant public sector controlled the 'commanding heights' of the economy was swept aside by reforms intended to free the private sector and integrate the Indian economy into world markets.

Since structural adjustment began growth has accelerated, as has the size of external trade in the economy. But India remains very much an agricultural economy. In 2000–1, primary production accounted for a quarter of value added but employed about two-thirds of the labour force.

As with other adjusting countries, India shares the task of reorienting government from its historic function of direct provider of goods and services to manager of the economy, and the tasks of enabling, regulating and contracting that this entails. This chapter explores the capacity of the Indian government to reform its role in agricultural trade. It overviews the organisational arrangements of government's interventions in agricultural trade, current reforms, performance of the arrangements and the forces for and against change.

Organisational arrangements

The modern history of food management operations in India goes back to the Bengal Famine of 1943 during which more than one million people died of starvation. The Food Policy Committee, formed in the aftermath of the crisis, advised that active intervention by central government in the management of food markets was required if such crises were to be avoided in the future. Tyagi (1990) argues that this period was a 'watershed' in the food policy of India in that it marked the beginning of an era of control which continues to this day.

The planning era which began in 1951 institutionalised this approach to the food economy. The basic rationale behind food policy in India was the need to stimulate domestic production and to ensure prices for basic commodities particularly grains did not rise. The trade policy for agriculture was highly restrictive and the domestic agricultural sector was insulated from world markets via central control over imports and exports. Although not formally part of the planning system, the food management system which emerged to govern the functioning of the domestic market was moulded by the same concerns. It is against this background that the current arrangements for agricultural marketing have to be understood.

Bureaucratic structure

Although agricultural marketing is a state function, the union government plays an important role in laying down the necessary policy framework. The Planning Commission determines the sectoral and regional allocation of resources and the Ministry of Finance is responsible for fiscal and monetary policies that have direct bearing on the food grains sector. In addition to these there are three union ministries that have a predominant role in the functioning of the agricultural marketing sector. They are the Ministry of Food, the Ministry of Agriculture and the Ministry of Civil Supplies, Consumer Affairs and Public Distribution.

The main objectives of food policy have been:

- to ensure remunerative prices to farmers
- to ensure the availability of cereals at affordable prices for consumers, and
- to hold a large buffer stock of food grains for national food security.

To achieve these goals, successive governments have pursued a strategy which divides the market for food grains, specifically rice and wheat, into two parts, one segment competitive and the other controlled. Food management operations by government in India involve complex procurement, storage, movement and public distribution mechanisms in the controlled sector and enabling and regulating of the competitive sector.

Controlled Sector

Procurement. The first stage of the food management network is procurement. This is the responsibility of the Food Corporation of India (FCI) which makes its purchases at fixed notified prices that are uniform across season and place. Prices for wheat and rice are based on recommendations made by the Commission for Agricultural Costs and Prices (CACP) and are intended to be remunerative to farmers. In the case of wheat, FCI makes its purchases through commission agents (*arhatiyas*) directly from farmers themselves who bring wheat to regulated markets and designated purchasing centres, and through state agencies acting on FCI's behalf. Sales to the FCI or its agents are on an entirely voluntary basis. The procurement arrangements for rice are somewhat different and depend on a levy applied to millers. Paddy, which is sold freely to millers by agents and brokers, is milled into rice and a fixed percentage of this must be sold to the FCI at stipulated prices that are intended just to cover procurement and milling costs. Once they have complied with their levy requirements, millers are free to sell the remaining rice in the open market at uncontrolled prices. The levy share has varied significantly over time and varies among the different rice-producing states. But it is argued to be a major limiting factor on profitability of rice milling.[2]

The scale of public procurement for both rice and wheat is very large and increasing. Although there are some marked inter-year fluctuations, especially for wheat, procurement has been rising over time. In 1994–5, 13.6 million tonnes of rice and 12.3 million tonnes of wheat were purchased by the public sector. These quantities accounted for 17 and 19 per cent of all-India production of rice and wheat respectively.[3]

While rice and wheat are grown in a large number of states, procurement operations are concentrated in a small number of surplus states. Just three states (Punjab, Haryana and Uttar Pradesh) regularly account for almost the entire procurement of wheat. Rice purchases are concentrated in six states: Punjab, Haryana, Uttar Pradesh, Andhra Pradesh, Tamil Nadu and Madhya Pradesh.

Purchases of rice and wheat are also concentrated with respect to time. In India there are two planting periods each year, the *kharif* season which runs from October to March, and the *rabi* season which runs from April to September. Wheat is a *rabi* crop while rice is grown in both the *rabi* and *kharif* seasons (the bulk of rice production takes place in the *kharif* season). *Rabi* crops are harvested in March and market arrivals begin in late March or early April. Market arrivals of *kharif* crops commence in mid-September. As a result of the pan-seasonal pricing policy in the controlled sector, market arrivals are concentrated into a very short period of time, placing great stress on the infrastructural facilities of both the public and private sectors. Under this system there is no incentive for farmers or traders to store grain.

Distribution. The primary purpose of the large-scale procurement operations of the FCI is to supply the Public Distribution System (PDS) which lies at the heart of the food security system of India. It is the responsibility of FCI to transport stocks of rice and wheat to central go-downs from where they are 'lifted' by the state/UT administrations and sold through Fair Price Shops (FPS). In the mid-1990s the number of FPS throughout India was estimated at over 400 000, each one catering to a population of about 2000 persons. Three-quarters of the shops are located in rural areas and the remainder are in urban centres (Ministry of Civil Supplies Annual Report 1996).

Food grains held in FCI depots are sold at a fixed price, the Central Issue Price (CIP), to the state/UT administrations. The CIP is insufficient to cover the economic costs of the procured grain and therefore a 'consumer subsidy' is incurred by central government. Retail prices at the FPS are usually higher than the CIP as state governments are permitted to add on state taxes and incidental costs to the price charged by FCI. There is considerable variability in the retail prices of different states, which means that access to grain by the poor is not uniform across the country.

Very large quantities of food grains are distributed through the PDS. During the period 1979–80 to 1994–5, public distribution ranged between 13.3 and 20.8 million tonnes per year. The share of PDS varied between 9 and 14.8 per cent of total production over the same period (Ministry of Food, and DES, DAC, Ministry of Agriculture).

A large proportion of PDS food grains are sold in the states of Kerala, Gujarat, Tamil Nadu, West Bengal and Andhra Pradesh. Those states with very large numbers of poor people, Orissa and Bihar, for example, account for a very small share of the distribution through the PDS. Overall control of the PDS rests with the administrations of the state/UT. State level Food and Civil Supplies Departments make monthly estimates of their requirements and make requests to the Central Ministry of Food. The Central Department of Food decides on the quantity to be allocated to each state and the FCI is informed of the relevant quantities.

Storage and stocks. The third element of the food management system is the storage and stocking of food grains. Stocks are held both for a national food reserve and to maintain steady availability through the public distribution system. The Technical Group on Management of the Bufferstock, which was constituted in 1975, recommended that a bufferstock of 12 million tonnes of food grains be held within the country, this quantity to be in addition to operational stocks which were set at a minimum of 3.5–3.8 million tonnes on 1 April and a maximum of 8.2–8.8 million tonnes on 1 July. In practice, stocks held have generally exceeded these norms with the bufferstock reaching a peak of 43 million tonnes in September 2001.

Competitive sector

Wheat and rice which are not procured by the FCI are sold in open competitive markets. While inter-state trade in rice and wheat is not physically restricted, there are nevertheless a number of regulatory and other interventions that affect the functioning of the domestic marketing network. Analytically, these can be separated into formal and informal restrictions. Formal restrictions include the Essential Commodities Act (ECA), 1956, which 'provides for regulation and control of production, distribution and pricing of commodities which are described as essential under the Act' (Government of India 1996: 13) Items in forty-four commodity categories are so labelled under this Act, including wheat and rice. Under Section 3 of the Act, if the government 'is of the opinion that it is necessary or expedient so to do for maintaining or increasing supplies of any essential commodity or for securing their equitable distribution and availability at fair prices ... it may, by order provide for regulating or prohibiting the production, supply and distribution thereof and trade and commerce therein' (quoted in Mooij 1995: 4).

The ECA is on the concurrent list, meaning that both central and state governments can issue orders under the Act. As state governments also have the authority to issue orders under the Act, there is tremendous variation in the control orders faced by different traders and different commodities in different states. The Act is justified on the grounds that the government must be equipped with the necessary powers to punish anti-social and anti-national activities such as hoarding and black-marketing. A major irony of the Act is that it creates the institutional environment which stimulates the very behaviour it was intended to deter. Opportunities for rent-seeking are expanded by the complex web of regulatory controls which the Act itself has created. There are said to be 225 control orders currently in force.

The Act has been amended from time to time to make it more effective. Major changes were added in 1981 making offences non-bailable and providing for summary trials in special courts (Essential Commodities (Special Provisions) Act 1981). These special provisions were extended to August 1997 on the basis of unanimous recommendation of the state/UT governments. Although traders and their associations lobbied vigorously, organising protests and demonstrations, they were unable to persuade Parliament to drop the special provisions. In 1995 alone, there were nearly 81 000 raids carried out, over 9000 individuals arrested, 2714 convictions, and Rs. 147.3 million value of goods confiscated under the Act (Government of India 1996: 13).

A second piece of legislation, the Prevention of Blackmarketing and Maintenance of Supplies of Essential Commodities Act, 1980, is also used to regulate private marketing activities. The Act, which was promulgated to prevent 'unethical trade practices' like hoarding and black-marketing, is administered under the state/UT governments and permits the detention of individuals if their activities are deemed to jeopardise the supplies of essential commodities.

Other legislation allowing restriction of trade includes the Prevention of Food Adulteration Act (PFA) affecting food processing and the State Agricultural Produce Markets Act, giving wide powers to licence. The Forward Contract (Regulation) Act, 1952, which established the statutory body known as the Forward Markets Commission, effectively prohibits futures trading in rice and wheat. Under the administrative control of the Ministry of Civil Supplies, the Commission advises the central government on forward markets, monitors forward and futures markets and makes recommendations to improve their functioning. At present, futures trading for six commodities is regulated through seven recognised associations. Forward/futures contracts are prohibited in respect of many essential commodities including food grains. 'Options' trading in these goods is also prohibited. Wholesale markets are confined to severely congested state-operated *mandis*, conditions causing excessive quality deterioration and wastage (World Bank 1999).

In addition to the formal restrictions embodied in the various Acts detailed above, a number of informal restrictions also 'regulate' the activities of the private sector. A prime example of these is credit controls which effectively prevent traders from gaining credit for stock holdings. Credit controls were first introduced in the 1960s when the government barred private trade from getting commercial bank credit for their operations. Although the situation has been relaxed somewhat since then, selective credit controls on inventory financing remain.[4]

A further example of informal restrictions is the priority given by Indian Railways to parastatals like the Food Corporation of India (FCI) in the allocation of rail wagons for transporting goods. As the cost of transport by road is significantly more than that by rail this discrimination actively deters the participation of private traders in the food grains market. Private traders who do use the road network to transport commodities are further handicapped by the numerous *octroi* tax points – the tax on movement of goods, which is a major source of revenue for many municipalities.

Finally, activity in the competitive food grains sector is affected by the operations of the controlled sector. The pan-seasonal and pan-territorial prices offered by the Food Corporation of India constrain the normal functioning of private trade because they distort the production and sales decisions of producers with respect to place and season.

In addition to regulating the competitive sector, the government also undertakes interventions to enable it to operate more efficiently. Examples of enabling activities include the promotion of standardised weights, measures, and grades by the Bureau of Indian Standards; grading and quality control (under the Directorate of Marketing and Inspection); and market information (provided by the Department of Agriculture and Cooperation in the Ministry of Agriculture). The last two are discussed in more detail in Chapters 9 and 10 below.

Reforms in marketing arrangements

Compared to the far-reaching liberalisation in the manufacturing and services sector, reforms in agricultural marketing have been tentative and limited to date. The main initiatives are listed in Table 3.1, distinguishing between reforms directed to internal and external trade.

There have been four main initiatives since 1991: to ease movement of agricultural products, to allow FCI to buy and sell wheat and rice on the open market, to allow exports of wheat, and to focus the PDS more towards the poor.

Easing movement of agricultural products. Various budget statements over the years (1993, 2001, 2002) have announced lifting of administrative restrictions on the movement of agricultural products within the country and abolition of stock limits on rice and wheat. On the face of it, the removal of administrative restrictions on domestic trade is a significant policy shift. However, it has been suggested that the rhetoric of change has not been matched by reality as the authorities have increased their reliance on other measures, such as the Criminal Procedure Code 144, under which District Magistrates can restrict movement of goods. Other local hindrance of private trade includes instructing Indian Railways to limit access to wagons by private traders hoping to ship wheat, as occurred in Punjab and Haryana in 1992, and implicit threats to revoke traders' licences if they tried to purchase from farmers (Pursell and Gulati 1995). Roadblocks at state borders can also be used to impede private traders attempting to move stocks by lorry.

Allowing FCI to buy and sell on the open market. The second major change was the entry of the Food Corporation of India (FCI) into open market for wheat and rice. This was not a liberalising reform, since its effect is rather to increase the activities of the controlled sector. Rather, the decision to authorise open market operations was prompted by operational and financial considerations. On the operational side, it was hoped that sales could be used to stabilise open market prices in the lean season. Financial imperatives arose because of the excessive carrying costs being incurred on large stocks. In 1993, when FCI was first permitted to sell wheat, existing stocks in FCI and hired facilities were at very high levels leaving insufficient storage capacity for the next season's procurement. High stocks add to the already considerable subsidies required by the FCI from central government. FCI has sold a growing tonnage of grains on the open market. But FCI stocks have continued to increase along with FCI's overall turnover.

Allowing exports of wheat. While basmati rice destined for export has been exempted from levy for many years, common rice has not. This changed

Table 3.1 Reforms: chronology of changes in marketing arrangements

Date	Internal market	External market
1991		Basmati rice for export exempted from levy.
Feb 1993	Zonal restrictions on movement of rice and wheat removed.	
May 1993		Non-basmati superfine rice for export exempted from levy and without quantitative or minimum export price restrictions.
Oct 1993	FCI authorised to sell wheat in the open market.	
Jan 1994	FCI authorised to sell rice in the open market.	
1994–5	1 Free inter-state movement of rice permitted without release certificates. 2 Central government advises states that stock limits for rice and wheat can be abolished	1 Exports of durum wheat allowed subject to quantitative ceiling and minimum export price (MEP later removed in Sept 1994). 2 Non-basmati fine rice export exempted from levy. 3 Small quantity of non-durum wheat freed for export without MEP. 4 FCI authorised to export non-durum wheat at a price not less than the central issue price.
1995–6		EXIM policy permits export of wheat products on open general licence subject to quantitative ceilings. Commitment to abolish quotas and reduce tariffs in line with WTO accession.
1997	Public Distribution System targeted more towards poor. Regulation restricting rice milling to small-scale enterprises repealed.	
2001–2	Commitment to remove restrictions on private storage and movement of grain and sugar	

in May 1993 when the government extended the levy exemption to (non-basmati) superfine rice for the 1993–4 *kharif* marketing season. This exemption was continued into subsequent seasons and further expanded to include fine rice. In 1994, the Ministry of Commerce proposed that a restricted quantity of high-quality durum wheat be exported. Agreement was obtained from the Ministries of Agriculture, Food and Civil Supplies, and export of wheat began subject to a quantitative ceiling of 300 000 tonnes for the crop year 1994–5 and a minimum export price (MEP) of US$160 per tonne (FOB price). The MEP was removed in September 1994. In subsequent years the ceiling has been raised and the wheat export market significantly expanded. However, other restrictions on exports remain. While rice destined for the export market is officially not subject to levy, state officials may nevertheless need a financial inducement to allow shipments across state borders. Since much of the export rice is grown in landlocked states, unofficial restrictions at state borders undermine progress in establishing an active export sector.

Targeting PDS to the poor. A revamped PDS was introduced in 1993, followed by the Targeted PDS in 1997, intended to target a larger food grain subsidy to so-called Below Poverty Line (BPL) households. The food grain ration to BPL households was to be raised from 10 kg to 20 kg per month and priced at about half of economic costs. Prices for Above Poverty Line (APL) households were to be raised to the full economic cost. But recent studies (Umali-Deininger and Deininger 2001: 325, Swaminathan 2000: 102) suggest that implementation problems under PDS, discussed below, have not been overcome by the reforms.

Overviewing these reforms, it can be noted that there has been no effort to overhaul the existing legislative structure which governs the marketing of food grains. This is true of statutes that segment the market and those that regulate its competitive part. This is in stark contrast to the changes that have occurred in the industrial and specifically the manufacturing sector where delicensing and dismantling of controls have been extensive. Further, the reforms that have occurred are not all in the direction of more open trade – notably the considerable expansion of FCI grain trading and storage.

The market-opening changes that have occurred in agricultural trade reflect the economy-wide pattern of reform, namely adjustments in the external sector. While some wheat exports are now allowed, policy continues to treat export markets as 'residual' markets, supplied only once domestic requirements have been met. This is also the principle that has guided the open market sales by the Food Corporation of India.

The approach to market-opening reform has been marked by caution, with minor adjustments introduced one at a time, partly to avoid subjecting low-income groups to sudden and large price increases for food staples, and partly to avoid confronting vested interests in the existing arrangements. A by-product of this caution is the flexibility it permits policy-makers as small changes can be reversed without exciting much comment or analysis. A major disadvantage is that this caution undermines the credibility of government commitment to a reform programme. But these reform initiatives appear to signal the direction of reform in the future.

Performance of the arrangements

Performance has both output and outcome dimensions. The output performance refers to the efficient accomplishment of marketing functions which is reflected in trends in production and the behaviour of food prices through time and location. The outcome dimension refers to the social welfare implications of these outputs. Successive governments in India have intervened extensively in food marketing arrangements in order to achieve both output and outcome objectives.

Outputs

Production and prices. One notable success of the food management operations in India has been the significant expansion in output of rice and wheat.[5] This has been achieved by a combination of policies including massive capital investment in irrigation, power, and research and extension, input subsidies, and a remunerative price policy to farmers to encourage them to expand production.

At the start of the planning period 1951–2 domestic rice production was 21.3 million tonnes and that of wheat 6.18 million tonnes. By 1960–1 these figures had risen to 34.6 and 11 million tonnes, by 1970–1 to 42.2 and 23.8 million tonnes, and by 1980–1 to 53.6 and 36.3 million tonnes. In 1994–5 rice production reached 81.2 million and wheat production 65.5 million tonnes respectively.[6]

Despite rapid population growth per capita availability of food grains was also rising over this period. Although there are some inter-year fluctuations, the broad trend of food grain availability has been upwards. In the early 1950s average annual availability of rice and wheat were 58 kg and 21 kg per capita, respectively. These figures rose to 70.3 and 37.8 kg in 1971, reaching 86.7 and 60 kg in 1995. Food grains production has outstripped population growth, enabling India to become entirely self-sufficient in rice and wheat production.

A related success is the decline in real prices of rice of wheat (Bhalla 1994). Indices of real wheat and rice prices show a fall from 100 in the base year 1970–1 to 65.2 and 78.9 respectively in 1991–2. Declining real food prices have a positive impact on consumers and net deficit food farmers and are welfare-improving.

Efficiency of the marketing network. There are many impediments to the efficient functioning of the marketing network, some of them natural and others imposed by government regulation, as outlined above. India is a geographically large country and the transport network in some regions is poorly developed and maintained, particularly in the remote and hilly areas. Nevertheless, it has been argued that, despite the hand-icaps it faces, the private market is able to operate without undue margins (Pursell and Gulati 1995), and that markets are regionally integrated to a degree, as indicated by covariance in prices (Palaskas and Harriss-White 1993). There is also evidence that private traders have been able to make food grains available at lower prices than the public distribution channels (Tyagi 1990). This reflects inefficiencies in public distribution as well as any efficiencies in private distribution achieved in spite of public regulations hindering private trade.

Outcomes

Preventing famines. Self-sufficiency in food production was accorded high priority by Indian planners as part of a two-pronged approach to eliminate famine. The second prong in this strategy was disaster relief, which recognised that famine can occur with no change in the physical availability of food if individuals suffer a loss of entitlement (Sen 1981). Effective mechanisms for scarcity relief have been in operation in India since independence. Under this system, financial responsibility for relief measures is shared on an equal basis between the centre and the states. Relief activities include providing employment for those able to work and free food for those not so able. In addition, extra food grain supplies are channelled through the public distribution system (PDS) during periods of stress.

The absence of any significant famine is testament to the success of the strategy adopted. India has successfully avoided famine despite experiencing significant falls in food production at various times in the last five decades. The most recent severe drought of 1987–8, which caused widespread crop failures in Rajasthan, Gujarat and Madhya Pradesh, was tackled with immediate implementation of a range of relief measures, including employment programmes and direct feeding schemes. Tyagi (1990) reports that at the height of the relief programme, 6.6 million person days of additional employment were being provided by the public sector.

Reducing chronic undernutrition. While the food management operations have helped to expand production and avert famine, chronic food insecurity remains a serious problem. Millions of people have inadequate diets to meet their daily needs. Despite its great costs to the Indian budget, the public distribution system (PDS) has been ineffective in reducing chronic undernutrition.

With the exception of Kerala, where it operates well, PDS fails to reach the poorest states and the poorest people. In Orissa, Bihar and Uttar Pradesh, which are states with very high incidence of poverty, PDS grains consumed as a percentage of total grains consumed were 0–3 per cent for the poorest fifth of the population, as calculated from the 1993–4 national sample survey of consumer expenditure. The equivalent figure for Kerala was 59 per cent, and for the richer states of Gujarat and Tamil Nadu 64 per cent and 60 per cent, respectively). The switch to targeted PDS in 1997 has resulted in much increased allocations to states with the most poverty (Umali-Deininger and Deininger 2001: 326). The impact of charging higher prices to the non-poor is not yet clear. But underlying problems in the system persist: irregular supplies and prices often exceeding the official price at fair price shops, inadequate storage making for large losses and low quality, and unaccounted-for leakage of grain into the open market where prices are higher estimated to reach up to 40 per cent (Umali-Deininger and Deininger 2001: 325 citing a study by Kriesel and Zaidi 1999; Swaminathan 2000: 102).

In sum, during periods of food scarcity the PDS does seem to be effective, but this ability of the bureaucracy to reach the poor during periods of stress is not matched at other times.

Subsidy costs. A second major failing of the current system is the enormous cost to the government of maintaining it. The rising cost of procurement and distribution has resulted in a growing and perhaps unsustainable subsidy. The subsidy required by the FCI can be measured by the economic cost of handling food items through the public system. The economic cost is defined as the procurement price plus incidental expenditure relating to procurement and distribution. Tyagi (1990) demonstrated that the economic cost had risen over the previous two decades. Furthermore, the economic costs have been rising faster than the procurement price, indicating that FCI's own costs of administering the system are a rising proportion of costs. In the early 1970s the economic cost was only 20–22 per cent higher than the procurement price; this rose to 30 per cent in the late 1970s, to 54 per cent by the mid-1980s and to 60 per cent now.[7] These rising costs underlie increases in prices to consumers at fair price shops and have meant that private trade has often been able to distribute with lower margins than the PDS.

Other impacts. Other impacts of the system are the distortion of incentives to farmers, encouraging overproduction of grains relative to other crops with higher value; and the underdeveloped state of private markets in grain (lack of forward markets, private storage, low value added in the food sector, low private investment in warehouses and cold stores) which encourages food grains to be marketed in seasonal peaks, with consequent destabilising effects on prices which then must be stabilised by state sales and purchases.[8]

Forces for and against liberalisation

Although the basic approach to the food economy pre-dates independence in 1947, the various reviews of the system since then have each endorsed a 'hands-on' approach to the management of the food economy. The approach originated from and has been reinforced by food crises. Critical to understanding the current system is an understanding of India's dependence on food imports which was regarded not only as undesirable but also as threatening in that it made India vulnerable to external meddling (e.g. United States' efforts to influence Indian domestic policy in the 1960s; see Joshi and Little 1994).

This led to self-sufficiency in food production becoming an overriding objective of Indian governments, articulated in Five-Year Plans, sectoral plans and national food strategies. It also led to a preoccupation with the need to manage scarce food supplies by means of a state apparatus despite the transformation of India's food economy to surplus production. This creates a powerful force against radical change of the arrangements – the belief that however inefficient and expensive they may be they deliver national food security.

Existing arrangements are reinforced by populism in Indian democratic politics. India is the world's largest democracy and state and central government officials must go to the electorate at least every five years. The political heritage of India, from its Ghandian and Nehruvian roots, involves a concern for the poor. A feature of the political scene is that all politicians aspire to be identified as the guardians of the poor.

Numerous groups gain enormous benefits from the existing marketing arrangements. Whatever the logic which justified the foundation of the existing system, as Krishnaji (1991: 189) argues: 'Once adopted, the policies lead … to the creation and strengthening of "interest groups" and to the transfer of policy making from the economic to the political realm.' In consequence, efforts to dismantle or alter current arrangements will trigger lobbying of political representatives by these groups. They include:

1 *Urban consumers.* Although there are some differences among the states, broadly speaking urban consumers gain a disproportionate share of the benefits of the public distribution system (PDS).
2 *Wealthy farmers and the 'middle peasants' in the surplus producing areas.* This group is anxious to maintain a system which guarantees to purchase its output at remunerative prices. Its demands are supported by politicians eager to secure votes.
3 *Parastatal personnel.* The FCI alone employs 70 000 persons all of whom enjoy protected public sector employment.
4 *Traders and millers.* While controls and levies irritate traders and millers, those able to circumvent controls and divert supplies to the open market gain large profits from the higher prices prevailing there.
5 *Bureaucrats.* The control system which characterises the Indian economy is heavily dependent on a large bureaucracy for its management and implementation. The bureaucracy have a vested interest in the maintenance of the 'permit raj', not only because it provides employment, but also because it provides them with opportunities to supplement their government pay through corrupt practices. It is debatable whether corruption in India is simply individualised rent-seeking or is institutionalised in that the behavioural norms are accepted and shared within society. As Mooij (1995) argues, the distinction is relevant to attempts to limit and eliminate corruption. If rent-seeking arises simply because individuals are optimisers then deregulation which removes the opportunities for rent-seeking will solve the problem. If, on the other hand, corruption is institutionalised then the solution is to break old norms and establish new modes of behaviour, otherwise system changes will simply involve pouring old wine into new bottles.

Compared to the entrenched and widepread support for existing arrangements, support for further liberalisation among the public is potential rather than actual. Consumers stand to benefit if a more developed food system provides a wider variety of better and cheaper food. The heavy weight of current subsidies on the federal treasury makes for ongoing pressure for reform from within government. Current efforts focus on transferring the administration and costs of public purchase and distribution of grain from the federal to the state level, but this is being resisted by the states.

Conclusion: India

In its food policy India has given highest priority to national food self-sufficiency. This has been secured by means of substantial subsidies to farmers and large-scale buying and selling by state agencies. However, the system contains great inefficiencies, distorts farm production towards food grains, has restricted development of private food trade (particularly storage) and infrastructure, and causes food grains to be marketed in seasonal peaks, with consequent destabilising effects on prices which then must be stabilised by state sales and purchases. Though it is effective in times of emergencies (e.g. drought), and targeting has been improved, inefficiencies mean the public distribution system is often ineffective in getting food to the poor in normal times.

Since structural adjustment began in 1991, the reform process has largely bypassed the agricultural marketing sector in India (Ahluwalia 1996). As Joshi and Little (1996) state: 'Much less has been done, and no framework of reform has been accepted, such as that provided by the Tax Reform Committee (TRC) in the case of industrial tariffs. It is notable indeed that the TRC ignored agricultural tariffs and trade. It implicitly assumed that agricultural trade would remain controlled.' This assumption is significant in that it highlights the pervasiveness of the belief that the status quo is unalterable.

While government is increasingly vocal regarding its intentions to reduce regulation of agricultural trade,[9] opposition from politicians, bureaucrats and traders gaining from the present system seems likely to continue to ensure that any reform is slow.

4
Sri Lanka

Marisol Smith and Frank Ellis[1]

Introduction

Sri Lanka is an island economy with an estimated population of 19.6 million people in 2001. The country has long been known for combining a low income per capita with good human development indicators. This reflects provision of a wide array of subsidised commodities and services since independence.

Sri Lanka has been fully independent since 1948 and has a functioning democracy. The main political parties are the left-of-centre Sri Lankan Freedom Party (SLFP) and the right-of-centre United National Party (UNP) which alternated power from the early 1950s until 1977 when the UNP entered a period of seventeen years of unbroken rule. In 1994 it was replaced by the coalition People's Alliance (PA).

A major feature of Sri Lanka since the 1970s has been civil and ethnic conflict. The roots of this can be traced back to 1956 when the incoming SLFP asserted the rights of the Sinhalese majority and made Sinhala the country's only official language. Economically there has been enormous dislocation in the north and east of the country with loss of agricultural and fisheries production, destruction of physical infrastructure and cessation of public services. The conflicts have absorbed a significant proportion of government expenditure, amounting to some 4–5 per cent of GDP.

1977 marks the beginning of Sri Lanka's adjustment period and distinguishes the country as an early adjuster. Sri Lanka therefore has long experience of adjustment of the state to new roles as it withdraws from its historical function as direct provider of goods and services. Here this facet of the adjustment process is explored in the context of agricultural marketing.

Organisational arrangements

Our purpose is to explain the origins and reforms of organisational arrangements in the agricultural marketing sector in Sri Lanka. The focus is on the marketing of the major staples in Sri Lanka, namely, rice and wheat, as these two commodities account for 30 per cent of average household expenditure and 50 per cent of average calorie consumption (Borsdorf 1993: xi).

Technical Features of Rice and Wheat Marketing[2]

Rice: production, storage, milling, wholesaling, retailing

Paddy production is the main crop of the non-plantation sector in Sri Lanka. There are two planting periods each year: the *Maha* season which runs from October to February and the *Yala* season which runs from June to September.

Paddy cultivation is predominantly small-scale with average farm size only five acres. About half of the paddy grown is retained on-farm, leading to small farm-gate sales. Much of the paddy sold is to 'collectors' (private traders and cooperatives) who assemble from a number of farmers for onward sale to millers. The paddy is sold at the farm-gate with the transport cost borne by the collector. Until 1996 paddy was also purchased by the government-owned Paddy Marketing Board (PMB) which bought paddy at a fixed guaranteed price known as the GPS. PMB bought at its stores, making most of its purchases from collectors.

Approximately 90 per cent of domestic rice consumption is met by domestic production of paddy. Given its highly seasonal production paddy must be stored for large periods of the year. Much of the rice consumed in Sri Lanka has been parboiled which reduces its storage qualities relative to raw rice and thus much of the storage is for paddy and not rice. Storage takes place at all levels of the marketing chain, including on-farm, at mill, and at traders' warehouses. As is the case for almost all agricultural commodities in Sri Lanka (see for example Menegay *et al.* 1995 on vegetables), paddy and rice are stored in large gunny bags with a capacity of 50–80 kg.

Returns to storage are highly dependent on seasonal price movements, costs of storage, and costs of finance. In the Sri Lankan case, the financial returns to storage have been highly variable. Estimates for the years 1991–4 suggest that storage was profitable in 1991–2, break-even in 1993, and unprofitable in 1994 (Harrison 1995: 3). These fluctuations arise in part because of the uneven participation of the Paddy Marketing Board in these years, and the significant reduction in the price of wheat flour in 1994, which resulted in a fall in demand for rice with a concomitant impact on rice prices.

Paddy is milled in a large number of mills across Sri Lanka. There are no recent estimates of the number of mills in the country, although a 1986 survey carried out by the Paddy Marketing Board concluded there were over 7000 mills. At the time less than 15 per cent of these were categorised as 'modern' establishments. The milling subsector is highly competitive with very small establishments servicing the farming community, milling paddy as required for on-farm consumption, and larger mills handling the immediate post-harvest output. Returns to milling activities have been estimated to be in the region of Rs 0.34/kg in 1993/4 (Harrison 1995: 5).

The Pettah market in Colombo is the main wholesale market for rice and other food commodities. In the mid-1990s there were an estimated seventy wholesalers competing there with around fourteen of them handling more than 30 tons per day. The real rate of return to wholesale activities was estimated at 18 per cent for operators owning their premises and 28 per cent for renter-occupiers (Harrison 1995: 6).

Rice is sold to consumers through a large number of diverse marketing outlets. In addition to small-scale traders selling in markets and shops, rice is sold through the retail outlets of the Multipurpose Cooperative Societies (MPCS).

Wheat: imports, milling, storage, retailing

Wheat is not produced in Sri Lanka and thus consumption of wheat and flour products must be met entirely through imports. Until 1968 Sri Lanka imported only wheat flour but since the establishment of domestic milling capacity, imports of wheat grain have become dominant. Since 1980 flour imports have been very small and in some years there have been no imports at all. Although there is significant inter-year variability, donor assistance has been important in financing wheat imports. Between 1980 and 1992, 45 per cent of total grain imports were donor-assisted, the largest share coming from American PL480 programmes (Borsdorf 1993). Until 1989 when the monopoly on wheat imports was transferred to the Cooperative Wholesale Establishment (CWE), only the Food Commissioner's Department (FCD) could import wheat. A limited quantity of private imports (250 000 tons) was permitted in 1994, amounting to approximately one-quarter of the total.

All wheat grain is milled by PRIMA Ceylon Ltd., a private company with a milling complex in Trincomalee. In 1980 PRIMA signed a twenty-five-year contract with the government of Sri Lanka with the latter guaranteeing to provide 435 000 tons of grain to be milled annually. Annual imports of wheat have exceeded this figure since 1981 and PRIMA has in practice milled all imports. There are no other wheat milling facilities in the country and new mills would find it difficult to compete because of the significant economies of scale afforded by the PRIMA mill which has a capacity of 3200 tons per day.

Wheat flour is stored by the Food Commissioner's Department (FCD) which is on contract to the Cooperative Wholesale Establishment to handle the flour milled by PRIMA. The FCD is the only wholesale distributor of flour in Sri Lanka and it sells to the Multi-Purpose Cooperative Societies.

Wheat flour is sold through the Multi-Purpose Cooperative Societies. Some three-quarters of all flour is purchased by bakers for the production of bread. The remainder is bought for household consumption or by private traders.

The case for state intervention

The particular circumstances of grain marketing in Sri Lanka provide little need for public sector intervention to prevent monopoly power. The risks of increasing returns to scale in storage, processing, transport, information and contracting leading to monopoly power seem small:

- In Sri Lanka paddy production is highly seasonal with two harvest periods of four–six weeks each year. The resulting concentration in marketed paddy requires large amounts of capital for storage and transport. This feature could lead to monopoly because it favours medium- to large-scale enterprises and lower unit costs. But in practice, because rice is a major food staple and as much as half of production is retained on-farm for domestic consumption, there is a tradition of on-farm storage capacity.
- Substantial on-farm storage reduces the pressure on storage downstream and therefore reduces the possibility of concentration at this point in the marketing chain. Storage is almost universally in gunny bags, a technology which is neutral in terms of enterprise size.
- The capital requirements for processing are sufficiently low to permit milling on a small scale. Indeed, given the uneven spatial and temporal demand for processing (farmers prefer to store paddy and have this milled as and when the need arises), small operators with low overheads are at a significant advantage because they can readily adapt to changing opportunities and combine processing with other activities, for example, farming.

- The transport of paddy and rice is done on small lorries or older passenger vans which have been converted into cargo vehicles. There appears to be no shortage of such vehicles and there is no need for specialist equipment for loading and unloading of lorries. Therefore, the transport sector is sufficiently competitive to avoid concentration in rice markets.

- Regarding transaction costs, the country is geographically small with few isolated areas which assists the spatial integration of the rice market. In addition, social overhead capital such as telecommunications and transport networks appear to be adequate in terms of linking all areas.[3] Whether government intervention is needed or not to facilitate price search, in the mid-1990s intervention via the now defunct Paddy Marketing Board (PMB) appeared at least to provide a guide. All participants were aware of prices at which the Paddy Marketing Board (PMB) was supposed to purchase paddy, even if the limited purchases by PMB meant sale to PMB was not possible. This price changed very infrequently and when it was altered the new price was widely broadcast via the radio and newspapers.

- Imperfections in financial markets may mean that producers find themselves embroiled in interlinked contracts with assemblers. There has been little recent research on the prevalence of interlinked markets in Sri Lanka, but farmers interviewed in the course of this research mentioned that collectors would provide loans in exchange for guaranteed sales. Thus, it is probable that in some instances collectors enjoy monopoly powers enabling them to generate significant rents. However, this problem is probably more marked in other crop markets where the government has been less actively supporting the provision of inputs.

While the above suggests there is no prima facie case for direct interventions in the paddy/rice marketing network, a partial case can be made for interventions in wheat imports and distribution. Wheat is not domestically produced in Sri Lanka and thus must be imported in the form of grain or flour. Grain imports must be milled and there are substantial economies of scale in the milling process. The privately owned PRIMA mill has a capacity of 3200 tons per day and is thus able to handle the entire wheat milling needs of the country, creating significant barriers to entry in the milling subsector. The nature of the contractual relationship between PRIMA and the government of Sri Lanka ensures that PRIMA is not guaranteed monopoly rents from its privileged position.

State intervention in rice marketing in Sri Lanka has been prompted by distributional and welfare objectives, including the desire to achieve national and household food security. Sri Lanka is nearly self-sufficient in rice production, with a self-sufficiency ratio averaging 90 per cent. Successive governments have actively pursued a policy of self-sufficiency on the grounds of national security and the Guaranteed Price Scheme (GPS) has been a central element of their approach.

The apparatus of interventions which exists in Sri Lanka can be understood in the context of the need to supply rice for the ration scheme which was operational from 1941–78. This scheme, which provided free or heavily subsidised rice to almost all households in the country, was fiscally unsustainable and was replaced in 1979 with a selectively targeted Food Stamp Programme. Under the current arrangements, households meeting the specified income criterion are entitled to receive food stamps which can be exchanged for specified food products in designated shops.

Household food security is closely linked to poverty and, as the poor spend a large share of their income on food, any policy which lowers the cost of food in real terms is welfare-improving for food purchasers. The most recent estimates available for Sri Lanka suggest that the incidence of poverty is still high, with a quarter of the population living in households with incomes below the poverty line (UNDP Human Development Report). While direct interventions into the marketing network are not necessary to provide consumption subsidies to the poor (and sometimes do not succeed in reducing consumer prices – see discussion of India above), nevertheless, the historical background of interventions in Sri Lanka is partly explained by a perceived need to lower food prices for the poor.

Public sector roles

Historically, agricultural policy in Sri Lanka has focused on three objectives, namely, enhancing paddy production, reducing reliance on imported rice, and improving the nutritional status of the population. To this end major policy instruments have included a rice rationing scheme, a producer support scheme (the Guaranteed Price Scheme, GPS), and a programme of import management.

The history of rice rationing in Sri Lanka stretches back to World War Two. The precedent for the Guaranteed Price Scheme (GPS) was also established in the war period. In 1942 an Internal Purchase Scheme (IPS) was introduced with the aim of supporting producer prices in a bid to enhance domestic production. The IPS was replaced in 1948 with the GPS by the post-independence government. The GPS operated as a price support scheme for producers and was intended to stimulate paddy production thereby reducing reliance on imported rice. In addition, the GPS acted as a mechanism for the procurement of sufficient paddy to supply the rice ration scheme.

During the years 1948–71 a number of different organisations had responsibility for the purchase and distribution of paddy under the GPS; in 1971 the newly established Paddy Marketing Board (PMB) was charged with this task. With periodic exceptions the GPS was above the import price (at official exchange rates) and therefore involved an element of subsidy to producers. Thus until the closure of PMB in 1997 the government operated a double subsidy, one for consumers and one for producers.

Import restrictions were also instrumental in the pursuit of self-sufficiency with a government monopoly on rice imports operating from 1942. This monopoly, which was first relaxed in 1988 and abolished in 1994, enabled successive governments to guarantee domestic producers protection from imports.

While rice marketing is almost entirely private the government retained some control. This control was exercised by the buying and selling activities of the Paddy Marketing Board, until its closure in 1997, and through the ability of the government to restrict imports and/or to change licensing and tariff procedures, including the terms on which the 'bondsmen' can release food security stocks on to the domestic market – with regard to volumes, quality specifications and minimum prices.

In addition, the rice sector has benefited from numerous other public interventions, including large-scale investments in irrigation infrastructure and subsidies on inputs, especially fertiliser. The scale of patronage to this sector is indicative of the fact that rice is regarded as central to every aspect of the Sri Lankan economy. Rice provides 40 per cent of total calorie intake, accounts for 25 per cent of consumption expenditure, and employs around half the labour force. To this extent the political fortunes of whatever group is in power are inextricably linked with movements in this sector. Protests by rice farmers over liberalisation of external trade in rice and cutting of subsidies to farm inputs highlight this link, and the dilemma of governments with limited resources trying to provide cheap food to the urban poor while sustaining a large population of small farmers.

In contrast to the marketing arrangements for rice, the marketing of wheat is largely government-controlled. Imports and distribution of wheat and flour are subject to a government monopoly, with the Ministry of Trade, Commerce and Food playing a central role, via its Cooperative

Wholesale Establishment (CWE) and Food Commissioner's Department (FCD). There are elements of private sector activity in the chain, however, as private transporters are used by the FCD and the monopoly milling company PRIMA is privately owned. In addition, the major end users of flour are private bakeries.

The public sector exercises further influence in the wheat market through two separate channels. Firstly, the price at which wheat flour is marketed domestically is subject to control by the government which has the power to instruct the Cooperative Wholesale Establishment (CWE) to sell to the Multi-Purpose Cooperative Societies at a specified rate. Until 1994 the fixed price permitted CWE to cover costs and make a modest profit on its wheat activities but the substantial price reduction introduced in August of that year resulted in enormous losses for the CWE. Secondly, the government has the ability to alter the tariff and licensing requirements for grain and flour imports.

Liberalisation reforms

Marketing arrangements for the major staples in Sri Lanka, and the public sector role in marketing, have altered very slowly over the last twenty-five years. Very little change took place before the end of 1989. Given that the adjustment process began in Sri Lanka in 1977, more than a decade passed without 'liberalisation' having any marked effects on the agricultural sector. The scale of the Paddy Marketing Board's operations had begun to decline in 1980, but this was not wholly, or even largely, due to an explicit policy choice with respect to the sector. Rather, it reflected the falling need for purchases under the Guaranteed Price Scheme which had traditionally been used as an instrument to ensure paddy supplies for the rice ration scheme. As a result of a budget crisis the universal rice ration was replaced with a targeted Food Stamp programme in 1979. This removal of 50 per cent of the population from the transfer programmes meant that the scale of procurement could be reduced. Thus, in 1978 PMB's purchases as a share of total paddy production was 36 per cent. By 1981 it had declined to less than 6 per cent (Gunawardana and Quilkey 1988: 18).

In the mid-1980s the Paddy Marketing Board's (PMB) operations were further reduced when it was obliged to divest a large part of its storage capacity and shed a large amount of labour. This was in part a result of an economy-wide drive to reduce the number of public sector employees and a more general drive towards administrative reform.

It was in 1990 that the biggest change in the marketing of rice occurred, namely the abolition of the Food Commissioner's Department monopoly on rice imports and establishment of a system of bonded warehourses for rice imports. Under this system, eleven trading companies were authorised to import and store rice without paying duty until the stocks are released into the domestic market. This arrangement between the private 'bondsmen' and the FCD permits the maintenance of a buffer stock of one month's consumption of rice without any significant budgetary costs being incurred by the FCD. Even this change proceeded very slowly with private imports subject to licensing restrictions until November 1994.

The changes in the wheat sector have been less marked compared to those occurring in rice. The private sector had almost no role in wheat grain imports with the exception of a brief experiment with private imports in 1994 when 250 000 tons were imported privately and then sold to the CWE. In the internal market, 1980 was a watershed in that it marked the commencement of the twenty-five-year milling mono-poly build-operate-transfer (BOT) contract between the government of Sri Lanka and PRIMA Ceylon (see Chapter 10). This arrangement introduced a substantial private sector component into the wheat sector.

Interventions in the internal market have also taken the form of price controls. Between the years 1979 and 1992, fixed prices for wheat flour were established for most points on the distribution channel. While this policy was abandoned in 1992 (leading to increased handling by private retailers), an implicit pricing policy remains in force with the Cooperative Wholesale Establishment (CWE) expected to price ex-mill wheat flour on a zero-profit basis. In 1994, a substantial subsidy for wheat flour was introduced with a price ceiling being imposed, leading to an approximate 30 per cent decline in flour prices and significant trading losses for the CWE.

Sector performance

The focus here is on performance in the rice market, but also with attention to the interaction between rice and wheat prices, and the significance of both these commodities for the nutritional status and food security of the Sri Lankan population.

The development of paddy production over the past two decades is an important backdrop to the discussion of marketing performance. Like other rice producing countries, Sri Lanka experienced rising yields per hectare and total paddy production with the adoption by farmers of high-yielding varieties in the 1970s (Borsdorf 1993; Department of Census and

Statistics, Paddy Statistics 1995). From the 1970s to the 1980s the average paddy yield for the country as a whole rose by roughly one ton per hectare from 2.0 tons to 3.0 tons. Since then average yields have levelled off in the range of 3.0 to 3.2 tons per hectare. Total output rose from 1.5 million tons in the 1970s to 2.3 million tons in the 1980s, and has subsequently averaged 2.6 million tons, enabling Sri Lanka to be more or less self-sufficient in rice during the 1990s. In the period 1993–5, rice output was more than required to meet domestic consumption, and the country was technically self-sufficient in this period.

Seasonal rice prices

A first measure of the performance of food crop marketing in Sri Lanka is the pattern of seasonal price formation in the rice market. A well-functioning marketing system would be expected to exhibit reasonable stability from year to year in the price spread between the lowest and highest price season, this spread to reflect competitive margins for crop storage given prevailing levels of interest rates, and these margins not widening over time without good reason. Conversely, a poorly func-tioning marketing system would display the opposite of these features: erratic and high price differences between low and high months, and perhaps a trend for these price differences to widen if the efficiency of marketing is deteriorating over time.

Results of the analysis of seasonal price movements for paddy in Sri Lanka between 1984 and 1995[4] demonstrate a marketing system which is functioning both effectively and efficiently with respect to inter-seasonal crop storage. There is little deviation of monthly prices from the average for the period. There is a maximum seasonal price spread between the lowest and highest price month of 12 per cent, and due to the lack of variation from year to year, this can be stated with an unusually high degree of reliability as being the seasonal margin for rice. This is not a high seasonal margin. The lowest price month is immediately after the main *Maha* harvest season in April. The highest price month is January before the new harvest starts. The price spread here is only 10 per cent over 4 months, or 2.4 per cent per month compound. This percentage describes a gross return to storage. When account is taken of the interest rate costs of funds tied up in stocks, plus the handling and deterioration costs of storage, it becomes highly improbable that uncompetitive profits are being made from seasonal storage in the Sri Lankan rice market. Moreover, the entire picture of seasonal rice prices presented here provides no evidence of monopolistic behaviour by traders and millers.

The conclusion of the seasonal price analysis is that the rice marketing system in Sri Lanka is competitive and works well. There is no evidence in the behaviour of long-term retail price series to suggest the contrary. The pattern of harvests means that the maximum storage period in the private sector is around four months, and this storage is probably undertaken both in paddy and rice, by farmers, traders and millers, mainly in relatively small quantities by numerous participants. Under such conditions it would be difficult for a single operator to 'corner the market', and highly risky for any individual actor in the system to attempt to make unusual gains by keeping stocks off the market. The evidence for Sri Lanka compares favourably with that for Java which has a considerably more developed infrastructure, and where a similar analysis to the one presented here found a gross seasonal margin of 11 per cent (Trotter 1992).

The producer–consumer margin

The producer–consumer margin (i.e. the difference between prices to consumers and producers) is a second output measure of the performance of the marketing system. This margin captures most of the factors in crop marketing which are not described by seasonal price movements. In the rice market it covers the costs and margins of purchase from farmers, paddy transport to mills, milling, rice transport, wholesaling and retailing. Again, as with the seasonal analysis, a well-functioning marketing system should exhibit stability in this margin, the absence of large fluctuations or trends in the size of the margin, and a share of the margin in the retail price which compares favourably with evidence of competitive markets elsewhere.

The producer–consumer margin for the rice market in Sri Lanka was examined, utilising monthly data for farm-gate and retail prices, for the period 1990–5. The producer–consumer margin displays remarkable stability over this period, all the more so given that this data is in current money terms unadjusted for inflation. This means that in real terms both the producer and consumer prices of rice declined during the first half of the 1990s. It is quite unusual, given seasonal price changes, for the proportionate share of the producer price in the retail price to remain the same month-by-month for a commodity like rice. Yet in the Sri Lanka case, the producer price is on average 76.5 per cent of the retail price, and the coefficient of variation around this average is only 3.5 per cent for the monthly price series over a six-year period. Thus the producer–consumer margin fluctuated in a minor way around a stable average of 23.5 per cent of the retail price in the early 1990s.

The analysis of producer–consumer margins reinforces the conclusion already reached that rice marketing in Sri Lanka is competitive and efficient. Were this not the case, it would be virtually impossible to observe price series and trends displaying these characteristics. While outcomes at the level of individual farmers, traders and millers are no doubt quite variable in practice on the ground, the combined effect of a diverse and heterogeneous marketing system is to produce extraordinarily homogeneous average price ratios and seasonal price relationships for the country as a whole.

Rice and wheat market interactions

Rice is the chief staple food commodity in Sri Lanka, but wheat (a wholly imported commodity in Sri Lanka, milled by the monopoly flour mill owned by PRIMA) plays a significant secondary role and the interactions between rice and wheat have critical policy relevance.

In summary, the rice market in Sri Lanka was characterised by the following features over the period from 1980 to 1995.

(a) domestic production grew, but not as fast as population growth, causing a potentially widening gap between output and consumption;
(b) the use of imports to close this gap tailed off, eventually leading to negligible imports and the rise in the self-sufficiency ratio;
(c) the gap was closed by a declining trend in per capita rice consumption, thus also resulting in a declining relative contribution of rice to dietary energy for the country as a whole.

Supply of wheat flour is determined by the volume of wheat grain that is imported annually. In summary, the wheat flour market in Sri Lanka was characterised by the following strong trends in the period 1980 to 1995:

(a) a sustained rise in wheat imports, surpassing one million tons for the first time in 1995;
(b) consequently, a sustained rise in the total and per capita availability of wheat flour over this period, with the latter reaching nearly 40kg per person per year by 1995;
(c) consequently, a rising trend in both the absolute and relative contribution of wheat flour to dietary energy, with the relative contribution increasing by 2 per cent per year in this period.

The above trends and interactions between the rice and wheat markets in Sri Lanka have not been the result of accidental market forces such as world price trends or switches in consumer preferences. The behaviour of both rice and wheat markets is strongly influenced by government food security policy, which has been substantially funded by donors in support of liberalising agricultural trade. Food aid in the form of wheat donated under a Multi-Year Food Assistance Plan between government and USAID has been sold into the market and the proceeds used to finance mainly the food stamp and refugee rehabilitation programmes.[5]

The results of the trade liberalisation programme are apparent in the changes to tariffs applying to rice and wheat imports. For rice, the *ad valorem* tariff varied between 25 per cent and 35 per cent during the first half of the 1990s. In 1995, it stood at 35 per cent. For wheat, an import duty of 25 per cent in the early 1990s was reduced to 20 per cent in 1992 and 1993, and was abolished in 1994. Both rice and wheat import volumes are also constrained by government decisions made through the Food Commissioner's Department (FCD). Although rice imports were fully liberalised in 1994, imports by the 'bondsmen' can only be released into the domestic market above a minimum price set by the Food Commissioner. A government monopoly on wheat imports remained in force despite the experiment with limited private imports in 1994.

The impact of policy is seen in price trends and fluctuations for rice and wheat, and in the relationship between them. These price movements in turn help to explain the strong trends already discussed in the physical volumes of rice and wheat consumed in Sri Lanka in the period under discussion. Though the price of both was falling, rice became significantly more expensive relative to wheat in Sri Lanka in the period under study, owing to the fall in price of imported wheat. The price of wheat flour at retail outlets is set by government. Actual retail prices may vary from the government fixed price, but not to any significant degree. A policy decision was taken in September 1994 to sharply reduce the price of wheat flour.

While policy decisions evidently resulted in the growth of wheat demand at the expense of rice, it is less clear whether this resulted from government controls over availability. Even if consumers have a strong preference for rice, the powerful price incentive to switch consumption from rice to wheat reinforces, even if it does not alone drive, the substitution in consumption that has occurred in recent years.

Impact on welfare of producers and consumers

The final aspect of sector performance concerns the outcome of these considerations for social welfare in Sri Lanka. Here, social welfare is used in its economic sense of relative impact on living standards in the large scale, and no attempt is made to distinguish the position of disadvantaged social groups defined by one criterion or another.

The prices of both rice and wheat declined in real terms over the period 1980–94 (i.e. relative to the prices of other goods and services consumed). This real price decline was sharper for wheat than for rice. Whereas for rice real prices moved unevenly and provide only an ambiguous indication of a slightly declining trend, for wheat there was a strongly defined decline in which real prices fell at roughly 4 per cent per year on trend.

Declining real food prices are good news for consumers, especially if sustained as a secular trend. It means that they can either buy more staple foods for the same income, or diversify their consumption basket while keeping up the same level of staple food purchases. In the Sri Lankan case, declining real food prices have especially meant the substitution of wheat for rice which could have long run implications on consumption patterns.

Declining real food prices are not such good news for staple food producers, in this case paddy farmers. While the trend in real rice prices may be ambiguous and even non-existent, paddy farmers are nevertheless potentially affected adversely by the market growth of ever cheaper wheat flour. This is because the market share of rice falls, possibly permanently, demand for rice decreases, and the direction of price pressure is for the real price of rice to fall too. When relative prices move as strongly as they did in 1994, when the wheat flour price dropped to Rs. 7/kg, it is clear that rice producers and traders will have faced powerful downward price pressures at all levels of the marketing system. A continuation of this trend may have implications for future rice production levels.

Conclusion: Sri Lanka

In the broad sense of responsiveness of actors (producers and consumers) to market developments, the food grains marketing sector in Sri Lanka appeared to perform well during the 1980s and early 1990s. It has been demonstrated, utilising time-series price data, that the private rice marketing system works efficiently and effectively.

Sri Lanka was one of the earliest developing economies to begin structural adjustment, following its financial crisis in 1978–9. The first major change affecting agricultural marketing resulting from the efforts to cut government expenditure was the ending of the universal rice ration and its replacement by a targeted food stamp programme. This reduced the need for public sector buying of rice and greatly diminished the Paddy Marketing Board's role. In 1988 it became impossible for the government to continue without external support and it accepted an 'Extended Structural Adjustment Programme'.

In the agricultural sector donor leverage under structural adjustment was exercised through the involvement of USAID. In 1990 a Multi-Year Food Assistance Plan was prepared and agreed between the government and USAID/Sri Lanka, which operated from 1992 to 1995. Under the programme local currency receipts from the sale of Title III food aid donations were used to support Sri Lanka's reform agenda. Thus wheat was sold to the Cooperative Wholesale Establishment at market prices and the local currency receipts from the donated wheat deposited in an account at the Central Bank. These counterpart funds were then available to finance activities agreed by the government and USAID. The programme operated a 'performance-based disbursement' system under which currency is released only if previously agreed benchmarks for reform had been achieved. The general thrust of the reform actions agreed under the Multi-Year Food Assistance Plan was the reduction of government control in the agricultural economy by reducing unprofitable government actions (particularly through CWE), increase in private firm's participation in the rice and wheat trade, and lowering of import tariffs. Overall, the Title III programme provided substantial budgetary assistance to the government. In 1994, these funds provided 30 and 40 per cent of budgetary allocations to the food stamp and rehabilitation programmes, respectively. In other words, there was a substantial financial incentive for the government of Sri Lanka to comply with the agreed reform benchmarks.

Nevertheless liberalisation proceeded only slowly. Certainly, the private sector has played an increasingly important role in rice importation in recent years with the government even willing to entrust private 'bondsmen' with food security stocks. But government's priorities were less with liberalisation than with continuing to secure welfare. Events in the mid-1990s highlighted some of the conflicts and costs in policy choice. It would appear that the incoming government tried to support simultaneously the two separate constituencies of rice farmers and poorer urban consumers, without heed to the interlocking nature of food markets between these two groups. The combined result of these two policies was to create a surplus of rice in the country and downward pressure on rice prices, causing financial losses for the PMB, the Cooperative Wholesale Establishment (CWE), and the bondsmen.

The attempted revitalisation of the near-moribund Paddy Marketing Board, by raising the price at which it was supposed to purchase paddy, conflicted directly with the downward pressure on rice prices induced by lowering retail wheat prices and encouraging consumers to switch from rice consumption.

Government's conflicting policy moves sent inconsistent signals to private participants via a number of channels. Firstly, tariff and tax structures for rice and wheat fluctuated widely and inconsistently. Secondly, there were ad hoc policy changes with respect to both rice and wheat, notably the simultaneous expansion in 1994 of the operations of the Paddy Marketing Board (PMB) and the introduction of a large wheat flour subsidy. Thirdly, while liberalisation of the policy environment proceeded there was no attempt to dismantle the apparatus of controls. Consequently, policy could be reversed with low transaction costs to the government. All these factors combined to have a disincentive effect on the private sector, which in the mid-1990s was understandably reluctant to invest in activities with large sunk costs, for example, storage infrastructure. The abandonment by government of PMB after 1996 – refusing it any further funding – brought greater clarity to policy.

In sum, the logic, direction and long-term impact of state interventions during the adjustment since 1980 has favoured the growth of the wheat market against the rice market. Consumers, but not rice producers, would appear to have benefited from these interventions due to falling real wheat flour prices and consequent downward pressure on real rice prices. Innovative means of providing food security (bonded rice imports) were developed. But by the mid-1990s the policy of cheapening wheat imports (using US food aid) to benefit the urban poor was reducing rice prices with potential negative impact on rural producers.

5
Ghana

Andrew Shepherd and Gideon Onumah[1]

Introduction

Instability has characterised the Ghanaian economy since independence in 1957. At that time Ghana was one of the most prosperous countries in Africa, with the highest per capita income in the region and very low inflation. Agriculture was the main source of wealth, making up about half of GDP. Cocoa provided about three-fifths of export earnings. Foreign reserves were strong as a result of buoyant cocoa exports and an abundant supply of labour for agriculture, much of it migrants from neighbouring countries. Disastrous economic policies in the 1960s and 1970s over-expanded the state and favoured import-substituting manufacturing and large-scale farming. By 1982 per capita incomes had fallen 30 per cent, export earnings were halved, and imports were at 30 per cent of their 1970 levels (Sarris and Shams 1991: 1, 3). In 1983, in the face of much internal opposition, the military government of Gerry Rawlings adopted an Economic Recovery Programme (ERP) financed by World Bank and IMF, to restore fiscal and monetary discipline and production incentives. This involved radical realignment of exchange rates and interest rates and cutting of government expenditure and employment.

UNDP's Human Development Indicators (HDI) (UNDP 2002) suggest that Ghana's economic decline, captured in the reduction of GDP per capita from \$1049 in 1960 to \$930 in 1991 did not have the severe impact on the quality of life which it might have done. Indeed, Ghana's overall HDIs show continuous improvement during 1975–2000. Population more than doubled, from 9.9 million in 1975 to 19.3 million in 2000. However, under-five malnutrition remains stubbornly high at 25 per cent. There are still large rural–urban gaps on most indicators, and rural poverty is much more substantial than urban. Employment in industry declined from 15 per cent in 1965 to 11 per cent in 1990–2, while the proportion of the population engaged in agriculture remains stagnant around the 60 per cent mark.

In 1993 Ghana made a partial return to multi-party democracy following elections in 1992. A feature of the 1992 elections was the considerably higher levels of public expenditure which were recorded in the period leading up to it. This is perhaps an indication of the fact that the benefits of structural adjustment in Ghana were extremely slow to emerge. As a result, the regime had to increase public spending in order to gain the support of the rural constituency. Increased public expenditure was apparently accepted by the IMF as a necessary price of democracy, despite the substantial degree of disruption to otherwise creditable progress in macroeconomic management. Ghana has since then completed two further presidential elections, including in 2000 replacing Gerry Rawlins, the architect of Ghana's structural adjustment reforms.

This overview of the changing role of government in Ghana's agricultural markets, sets out the characteristics of Ghana's agricultural production and markets subject to the limitations of data quantity,[2] and the changes in the role of the state.

Organisational arrangements and performance

Ghana has seen major changes in quantity, location and nature of production of major commodities since the 1970s. The collapse of the Ghanaian economy during the 1970s and early 1980s led to significant loss of agricultural markets both national and international. This was most noticeable in the export sector where cocoa exports fell as a result of production declines. At the same time food crop production increased, despite falling prices and increasing imports. Non-traditional agricultural exports showed only minor increases. Among these the significant growth areas were in cotton and sheanuts. Overall, the share of agricultural exports in total exports fell in the 1990s, and that of non-agricultural exports increased – particularly gold, timber, diamonds, manganese and electricity. Manufacturing exports stagnated (ISSER, 1995: 68, 103).

The focus here is on cocoa, Ghana's main export crop, and on the principal food crops maize and rice, and non-traditional agricultural exports.

Cocoa

The Ghana Cocoa Board, better known as Cocobod, has dominated Ghana's cocoa industry for most of its history. Originally set up as a profit-making body to buy, process and market cocoa, it was made into a statutory corporation in 1947 and became 'more like a ministry' (Sarris and Shams 1991: 134), employing over 60 000 people in the mid-1980s. Besides cocoa, Cocobod also purchases and exports sheanuts and coffee. In the post-independence decades before structural adjustment in 1983, Cocobod suffered from the capture of its stabilisation fund by government and from overvaluation of the Ghanaian cedi (ibid.). Despite severe downsizing in the 1980s, Cocobod in 2000 remained the driving force in the sector. It consists of a number of interlinked organisations.[3]

Internal trade in cocoa from the 1930s to 1961 was carried out by private buyers. By the 1960s the trade was dominated by about twenty private companies, mainly of European origin, which were also engaged in the importation and distribution of consumer goods throughout the country. In 1961 the state-funded United Ghana Farmers Co-operative Council was given a monopsony on internal trade in cocoa, largely for political reasons.

A partial competitive arrangement in the internal marketing system was introduced in 1966 with the licensing of sixteen buying agents by the then Ghana Cocoa Marketing Board. This was to enhance efficiency in the internal trade. But by 1972 licences for many small Licensed Buying Agents (LBAs) were withdrawn because of indebtedness to the GCMB and lack of capacity. Between 1973 and 1976, two of the eight remaining companies accounted for about 94 per cent of the total volume of crop purchased internally, with another three handling 5 per cent and the rest accounting for only 1 per cent (Kotey *et al.* 1974).

A return to the state monopsony buying arrangement was made in 1977 with the establishment of the Produce Buying Company (PBC), a subsidiary of Cocobod, as the only buyer of cocoa in the country. The change was supposed to address problems of delayed payments to farmers due to diversion of funds by LBAs and over-declaration of purchases which created problems with external trade in cocoa (Nyanteng 1995).

Consequently, from 1977 until the domestic marketing of cocoa was liberalised in 1992, the state controlled purchasing, storage, transporting, quality control, pricing, finance and exporting of the crop. These functions were carried out through Cocobod and its subsidiaries. The infrastructure required for the trade, including storage sheds, equipment for weighing and quality control, bulk haulage facilities, feeder roads to producing areas and port handling facilities, were owned and maintained by the state. Staff were recruited to handle all these operations with only marginal private sector involvement in the trade, mainly in the contracting-out of bulk haulage in emergency situations.

In addition, Cocobod owned and operated cocoa and coffee plantations, produced and sold insecticides to farmers at a subsidy, provided medical facilities to staff and farmers and supplied consumer goods to farmers at government-controlled prices.

When structural adjustment began in 1983 dissatisfaction with the low price paid to farmers and Cocobod's high costs (particularly in the Produce Buying Company, buying cocoa from farmers) focused attention of reformers on the cocoa trade. Reforms since then include drastic cutting of Cocobod staff; removing restrictions on private trade in coffee and sheanuts; licensing of private cocoa buying companies (1992) which now account for about 30 per cent of purchases from farmers, floating of a majority share in PBC to producers and private investors in 1999 (which was badly undersubscribed); opening of 30 per cent of the export market to licensed exporters in 1999 (for which they were not yet sufficiently prepared in the 2000–1 season); merging of Cocobod's extension service into the government's service; and agreeing to give the same access to credit and warehousing to the licensed buying companies as it gives to PBC.

While the World Bank and IMF have pressed for full liberalisation of both external and internal trade in cocoa, government has chosen to move slowly.[4] Despite the inefficiencies in the domestic trade, Cocobod's good overall performance provides government with justification for a measured pace of reform, in order to minimise risks to farmers. Ghana's cocoa exports are premium quality, sustained by the Quality Control Department. Cocobod's Cocoa Marketing Company has a high reputation with its foreign customers, and a strong credit rating which enables it to raise finance on international markets for buying the crop. Cocobod has also demonstrated an ability to manage change, as its working with the licensed private buyers in the domestic market demonstrates. Reforms are still at an early stage, since private firms have only some 30 per cent of domestic and external trade, and the eventual role for Cocobod is seen to be as a regulatory and promotional organisation. Reform of Cocobod is discussed in more detail in Chapter 9.

Performance

Cocoa exports declined from a high of 557 000 MT in 1964/5 (one-third of world production) to a low of 158 000 MT in 1983/4. The factors in this decline have been much debated and include a long-term trend of declining world prices accompanying increased production (from about 1.2 billion tonnes to 2.4 billion tonnes from the 1960s to mid-1990s), greatly increased planting in Indonesia and Malaysia, steady expansion in Côte d'Ivoire and Brazil, the overvalued exchange rate in Ghana, which made investment in cocoa unattractive, and the loss of Ghana's cheap labour supply from Burkina Faso to Côte d'Ivoire (Kofi 1993: 32).

Devaluation of the cedi in the mid-1980s, plus producer price increases, resulted in some recovery of production. Whereas the producer price of cocoa was so low that considerable amounts were smuggled to Côte d'Ivoire through the 1970s, by the 1990s Ivoirian cocoa was sold in Ghana. However, the producer price as a proportion of the world price was still far behind that received by producers in Malaysia or Nigeria (LMC 1996b: 1).

After rises in producer prices and production in the late 1980s, the failure to raise real producer prices in the first half of the 1990s was partly political: a substantial proportion of the difference between world prices (f.o.b. prices) and producer prices is made up of government export tax, which was the government's biggest single source of revenue. When introduction of VAT in 1995 was abandoned after violent protests, there was renewed pressure from the Ministry of Finance to increase cocoa export tax levels. But with VAT in place from 1998, the commitment was made to raise the producers' share of f.o.b. prices to 67 per cent, with a production target of 500 000 tonnes by 2004 and 700 000 tonnes by 2010. Despite the glut of cocoa in 2000 causing falls in world prices, producer prices in Ghana were maintained, bringing them up to some 70 per cent of f.o.b. prices. It is intended that production increases in the long run will result in revenue increases to offset short-term public costs of the policy.

Critics (World Bank, IMF) argue that this level of price support is now too high, and government should instead be using its resources to improve infrastructure for farmers, upgrade feeder roads to marginal areas to attract private buyers, reduce taxes on exports and encourage value-added processing.[5] The reputation of Cocobod for high quality cocoa[6] strengthens government's case for gradual reform. Yet the strategy of increasing production, on which government is now set, raises the question whether government has the capacity to reduce the negative effects of expanded production on the long-term welfare of Ghanaians.

Should Ghana increase cocoa production?

Expanding cocoa production is perhaps the single easiest way to promote economic growth, since the cocoa culture is so firmly established in Ghana. Its return to the 300 000 ton mark in the 1990s was achieved partly by the reduction or elimination of smuggling to Côte d'Ivoire, as prices in

Ghana have improved, partly due to better care of existing farms (the short-term response to price), partly by the development of new farms at the expense of forest, and partly by replanting old, diseased farms.

But Ruf (1995: 198) estimated that destruction of the forest in the Western Region accounted for the major part of the expansion since 1980. Yields from newly developed farms are certainly substantially greater than on replanted farms, so the incentives are there to develop them. Wage costs are generally lower in forest areas due to migration. The degree to which an improvement in producer price leads to increased replanting and better care of existing plantations compared to new planting in the forest varies. It depends on whether the necessary infrastructure for replanting is in place, whether new forest land is accessible (it currently is in Ghana), whether access to land in areas for replanting is orderly or characterised by conflict and dispute, and whether credit and subsidies are available for replanting, either by liberalisation or by a reduced level of taxation, combined with restrictions on forest destruction and facilitation of cocoa replanting. Road and other market infrastructure investments (information, purchasing centres etc.) would need to complement the replanting focus. Without such policies and government capacity to implement them, which does not currently exist in any comprehensive way, price incentives will simply lead to forest destruction. Furthermore, they will maintain Ghana's dependence on cocoa as its leading agricultural product. It could be argued that increased coffee production and the search for a better balance of exportable and industrial crops would be advantageous. In order to achieve this, maintaining cocoa taxation levels above the 'optimal' level may be desirable, in order to fund forest protection and cocoa farm rehabilitation.

Finally, significantly increased production would probably have a depressive effect on already weak world prices, especially since a high proportion (about 80 per cent) of Ghana's produce goes into the high-quality market. The experience of cocoa-producing countries is that it is very difficult to prevent prices falling when supply exceeds demand.

Food crops and trade

There has been a sustained upward trend in food crop production in Ghana from the 1970s, despite decreases in real prices, removal of subsidies on farm inputs in the 1980s, and variations in the inter-sectoral terms of trade between food crops and cocoa – which favoured food crops until the late 1980s, then cocoa thereafter (Nyanteng 1980; Commander *et al.* 1989; Ofosu 1995). Causes of the trend are not certain. On the demand side the long-term decline in prices of domestic staple food crops can possibly be attributed to slow growth in demand, owing to the transfer of tastes to rice (partly imported) and wheat (which is not grown in Ghana); the declining value of rural labour; the stagnation of urban incomes; and a slowing of urbanisation. Production seems to have easily kept pace with growth in demand for all commodities except rice. On the supply side, producing food crops may have been essential for household food security during Ghana's economic disruption in the 1970s and 1980s, while commercial motivation may have been more important with growth of incomes in the 1990s.

State intervention in food supply and distribution has been slight and unsustained. It was confined to the setting up and subsequent abandonment of the Ghana Food Distribution Corporation in an abortive attempt to provide guaranteed minimum prices for maize and rice, taxes on external trade in food (principally on rice imports), and the inconsistent use of food aid imports in the late 1980s and early 1990s.

Maize

Maize is among the major Ghanaian staples and along with other grains accounts for about 60 per cent of the calorie supply of rural households and 50 per cent of that of urban households (World Bank 1992). Production is concentrated in the middle belt of Ghana, comprising Ashanti, Brong Ahafo and eastern regions, which account for about 75 per cent of total annual output (Coulter and Asante 1993). It is grown mainly as a cash crop in these regions where it is not an important staple. Consumption is concentrated in the southern regions and as a consequence the main trade flow is southwards though there are also significant flows northwards, albeit via informal channels, to the Sahelian nations of Burkina Faso, Niger and Mali (ibid.).

The maize trade is characterised as atomistic with most participants being small traders (Coulter and Asante 1993). The main marketing function involves spatial movement with rather low levels of value addition through processing. Like most food crops the bulk of marketed maize reaches the consumer in unprocessed form (Dapaah estimates that 90 per cent of food crops reach the consumer in unprocessed form in Ghana – Dapaah 1993). Though seasonality of production and wide intra-season price variability offer considerable scope for profitable storage, traders normally do not store much grain, apparently because working capital limitations make quick stock turnover a more critical objective (Coulter and Asante 1993). On-farm storage is more common though the traditional storage system is usually thought to lead to substantial losses (Nyanteng and Dapaah 1993).

There is varying empirical evidence as to whether the market is well-integrated and competitive.[7] Formal studies apart, the general perception in the mid-1990s was that while the retail trade is reasonably efficient, the wholesale grain marketing subsector is weak and inefficient (World Bank 1992). This has been attributed to lack of suitable storage and transport facilities, poor road and market infrastructure (World Bank 1993) and limited availability of credit (Coulter and Asante 1993). Weak wholesale maize marketing is seen as a major factor in the loss of Ghana's competitiveness in the production of maize (Dapaah 1993; World Bank 1992). It is also blamed for the failure of substantial exports to take place, particularly to Sahelian West Africa and other parts of Central Africa (Coulter and Asante 1993).

Unlike cocoa marketing reforms, which since the early 1980s received primary focus in the government's structural adjustment programme, maize marketing appeared on the reform agenda only in 1990. This might be related to the fact that change in maize marketing was not among the main conditionalities demanded by donors. Maize marketing reform involved the abolition in 1990 of the guaranteed minimum prices (GMP) offered via the state-owned Ghana Food Distribution Corporation (GFDC). The Ministry of Food and Agriculture also began to provide, via national media, market price information on major food crops including maize (see Chapter 12). But ending the GMP was a nominal reform only (see Chapter 9) since at no point had the GFDC accounted for more than 10 per cent of the maize trade, and was down to about 1 per cent by the early 1990s. Furthermore, these reforms proved woefully insufficient in dealing with the underlying weakness in the wholesale maize trading subsector caused by poor infrastructure and limited access to credit.

Rice

The big production increases of the 1970s slid back by the end of the decade. In the 1960s and 1970s Ghana had a policy of agricultural import substitution, which generated substantial investments in rice farming in particular, since rice was a major import which could be grown in Ghana. By 1979 the banking system had become severely overstretched by defaulting rainfed rice farmers. Credit collapsed and the large farms established by politicians, army officers and business-men largely in northern Ghana shrank or disappeared. Subsidised bank finance underlay the entire rice production boom (Shepherd 1979). It came to an end once banks were forced to exercise some discipline under the IMF stabilisation programme (1983–6).

As a result of this harsh experience, banks were disinclined to lend to agriculture in the mid-1990s. Rice had become a small farmer's crop, though there were still some big farmers in it. Much of the increase in production is accounted for by irrigated production in the rehabilitated Dahwenya scheme (eastern region) and other irrigation schemes in the northern and upper east regions. Government offered guaranteed minimum prices for rice until 1990, but did not intervene adequately in the market to assure a minimum price to many. As the cedi devalued, imports rose in price, so domestic production (largely in the north) was beginning to compete with imports into the south again. There were significant food aid imports of rice until the mid-1990s. The combination of these, commercial imports and the depressed credit market were probably sufficient to restrain production growth. The ending of food aid imports coupled with price rises made rice production a better prospect. The risks of rice farming had also been substantially reduced by the MFA's development of a low-risk rice cultivation system, involving the construction of contour bunds to hold water. After 1994 rice imports sharply decreased while cereals production remained on an upward trend.[8] Tariffs and sales tax on imported rice amounted to 40 per cent, giving locally produced rice some protection.

Food aid

Food aid constituted on average 24 per cent of all rice imports between 1989 and 1993. It also constituted 37 per cent of all wheat imports. USAID began to use rice as food aid in 1992, alongside food aid in wheat. Rice food aid was technically justified by the World Bank's calculation that the Domestic Resource Cost (DRC) of rice was > 1, which implied that its production was uneconomic in Ghana. Other interests in food aid imports were those of revenue generation from the counterpart funds (Ministry of Finance/Customs and Excise). In the food aid committee these interests appeared to dominate considerations of not upsetting local markets when it was being decided when to auction the rice. In 1994 imports were selling at lower prices than Ghanaian rice and both large and small rice traders complained of being unable to compete with imported rice.[9] USAID stopped its Title III food aid imports to Ghana in 1995.

Agricultural non-traditional exports

Sales of coffee and sheanuts have increased dramatically since the export market was liberalised. Traders are allowed to export, determine their own prices, and Cocobod's role in trade has ceased. Since it takes several years before new coffee bushes produce coffee, most of the increase in coffee marketed post-liberalisation must have come from imports from Côte d'Ivoire. However, it was anticipated in the mid-1990s that the then better international prices for coffee, combined with competition for limited production in Ghana, would encourage farmers to increase planting. Cocobod produced 10 000 coffee seedlings per year all of which were bought by farmers.

In the mid-1990s competition in both the coffee and sheanut markets was intense, with farmers being offered higher prices for their produce, to the extent that prices for both commodities have reached close to world market levels and buyers have reported losses; competition was also expressed through buying unhulled rather than hulled coffee to help farmers, and by giving prizes and bonuses to loyal farmers; exporters have also competed by bulking exports, and by getting contracts with overseas buyers including access to credit at international rates of interest (about 8 per cent p.a. compared to about 38 per cent in Ghana). But the increases in exports of these, as well as pineapple, vegetables, cashew nuts and bananas were offset by declines in other agricultural non-traditional exports, e.g. fish, coconut products. This raises the question whether successful stimulation of non-traditional exports by liberalisation was confined to products in which no further enabling role by government was needed.

Conclusion: Ghana

The Ghanaian state's capacity to maximise development gains from the liberalisation of agricultural markets, and to minimise losses, has been limited and unevenly applied.

There has been little improvement in civil service capacity. The structural adjustment reforms 'were designed without properly assessing the capabilities and "state of readiness" of the main implementing agency, the civil service' (Larbi 1995). A civil service reform programme was designed largely to improve its performance in meeting the various conditionalities imposed by the structural adjustment loan agreements. It concentrated on the ministries which are crucial players in the wider structural adjustment process. But other priorities of the government, such as decentralisation and restructuring of ministries, were not included in the structural adjustment loan conditionalities and have made less progress.

The cocoa sector illustrates the various capacity issues. Government has given high priority to cocoa industry affairs. While the initiative for policy reform has often come from the World Bank, government has generally responded positively, acted on agreements made under loan conditions, and has been able to resist demands for rapid liberalisation where it feels a gradual process is more beneficial. However, a concern is that policy formulation for the cocoa sector is divided between Cocobod and the president's office, with the Ministry of Agriculture supposedly involved, but very much the weakest player. This means that the links between cocoa policy and agricultural policy in general may be weak. Given the historical importance of cocoa farming in the destruction of Ghana's forests, there is also a clear policy role for the Ministry of Environment and Forests. However this ministry does not appear to be part of the cocoa policy-making network. In addition there is concern about the adequacy of the technical and economic models used to guide policy, limited data, and a dearth of assessment of policy impacts.

Government's medium-term strategy for the cocoa industry in 1999 bears out these concerns.[10] While it articulates a clear programme for increased production and private participation in the industry, it does not emphasise limitation of damage to forests as a result of stronger incentives to extend cocoa farming.

The food sector has not benefited from the degree of government attention given to cocoa, perhaps because donor conditionality focused almost exclusively on the export crop sector. The privatisation programme has proceeded slowly; with the result that food marketing is still characterised by uneven and sometimes inconsistent state intervention, as in the sudden decision to allow GNPA to become a major player in rice imports and domestic purchases in 1995.

Government capacity to provide and maintain physical infrastructure in rural areas remains low, leaving many areas inaccessible and less able to benefit from liberalisation and competition in agricultural markets. The centralisation of government decision-making in the ministries in Accra hampers experimentation with a greater range of arrangements for investment in infrastructure. It is at local level that capacity needs to be created most urgently for infrastructure development. Government's decentralisation programme transfers 5 per cent of tax revenues to the District Assemblies Common Fund for development purposes; but this is far from adequate to meet the problem.

Government's limited capacity in public financial management and facilitating banking system development also constrains realisation of benefits from liberalisation of agricultural markets. In recent times the flow of credit to the private sector has been limited by high government debt. Financing the government's deficit has absorbed domestic credit and resulted in tight controls on further credit expansion, in order to curb inflation and excessive devaluation. This credit constraint is important in that it limits the extent to which Ghanaian traders are able to play a role in liberalised markets. The contrast with Cocobod's access to finance is stark. Cocobod has been able to use its strong credit rating to raise finance for crop purchasing from international finance houses, which can then be passed on to producers via PBC and the licensed buying companies. The only other agricultural commodities which have benefited from a reasonable flow of credit, and that only from one state bank, have been cotton and poultry. In the case of cotton the dominance of one previously state-owned company and the tying of producers to processors through advances and contracts make repayment relatively straightforward.

In general, the Ghanaian banking system still copes poorly with the fact that the agricultural economy is characterised by smallholdings. Even where small farmers come together, as in the case of Kuapa Kokoo, and are guaranteed by an international NGO, banks are still very cautious in their lending. Ironically, the collapse of the banking system in the 1970s was caused not by the defaulting of smallholders, but by defaults of big borrowers supported by the state (e.g. in mechanised rice farming in the north).

6
Zimbabwe[1]

Introduction

The role of government in agricultural markets in Zimbabwe has under-gone successive radical changes, driven by a wholesale shift in political power in 1980, and a continuing macroeconomic crisis. The context is one of decline in living standards and public services from relatively high levels compared to other African countries. Per capita incomes in Zimbabwe are now lower than in the early 1970s. It is one of only five countries whose Human Development Index rating is now lower than in 1980 (UNDP 2001 'Human Development Report'). Its health delivery system is ranked bottom in a survey of 191 countries by the World Health Organisation.[2]

Zimbabwe had a proud record in crop and livestock research, seed variety development, tsetse control, agricultural extension and market-ing, with its achievement after independence of greatly increased production from communal areas hailed worldwide.

At the time of writing (2002) the occupation of commercial farms by government-backed squatters leaves uncertain the future role of government in agriculture. In the most extreme scenario (collapse of commercial and export agriculture, massive expansion of subsis-tence farming) a need for radical increase in public agricultural services would confront a future government, if incomes from agriculture were to be raised. Funding and managing such services would be a formidable challenge and they could not be confident of success, as the lack of productivity increases on resettlement schemes from the 1980s forewarns.

Organisational arrangements and reform

In Zimbabwe agricultural services are provided by government through the Ministry of Lands and Agriculture (MOLA), by farmers' associations, by NGOs and by private firms (some of the main firms being former commodity parastatals, e.g. for cotton and dairy products. MOLA provides a variety of services:

- agricultural policy analysis and formulation (Policy Division)
- disease control and quality assurance (innoculation of livestock in communal areas, foot and mouth disease control, combating tsetse fly, meat inspection, phytosanitary controls at borders) (Department of Veterinary Services)
- food price assurance via a strategic grain reserve and control of import and export of grains using a permit system (Policy Division operating with Grain Marketing Board)
- laboratory services, agricultural research and seed development and regulation, milk inspection (Research Division)
- technical and extension advice and services, including irrigation and land (Agritex).

Zimbabwe has a strong network of farmers' associations, organised by activity (e.g. grains, tobacco, meat) within the Commercial Farmers Union (CFU), and as lobbies for particular groups of farmers, notably the Zimbabwe Farmers Union (ZFU) for communal farmers, and a black commercial farmers union. The latter emerged in the 1990s, have little organisational reach and reflect above all the racial divide within the farming sector, since the CFU is dominated by white commercial farmers. The CFU invests in research and was instrumental in setting up the Zimbabwe Agricultural Commodities Exchange (ZIMACE). The ZFU set up a pilot market information project in 1996.

The experience of Zimbabwe demonstrates the primacy of politics in determining the role and performance of government. Successive governments in Zimbabwe have intervened substantially in agriculture, with different redistributional purpose. Three broad phases of agricultural policy can be distinguished, following Jayne and Jones (1997).

Colonial phase

In the colonial phase, after the establishment of substantial settler farming in the 1920s, government intervention was designed to assist settlers. Maize, tobacco, beef and milk were the key commodities.

Monopoly state marketing channels (marketing boards) were the method chosen for maize, beef and milk, while tobacco marketing was coordinated through regulations. These were set up from the depressed early 1930s in line with the worldwide tendency at the time to put in place protectionist, state-dominated agricultural support structures to assist farmers in the depression. Fixed prices (pan-seasonal and pan-territorial) were paid for maize delivered to the state marketing boards. Lower prices were paid by marketing agents in the African areas (the Tribal Trust Lands). Through the controls on movement of grain and animals a few licensed mills and the parastatal Cold Storage Commission dominated urban flour and beef supplies. Rapid industrialisation and urbanisation in the region in the 1940s–1960s provided a growing market base. Commonwealth preferences assisted beef exports. Hybrid maize varieties developed by national agricultural research were widely adopted, yielding productivity increases on commercial farms (Eicher 1995). The onset of sanctions after UDI in 1965 further oriented policy towards self-sufficiency and marginalisation of private trade, with the extending of state marketing controls over other crops including soybean, coffee, wheat and cotton. 'The agricultural marketing system was designed to service large-scale commercial farmers and to achieve self-sufficiency in foodstuffs and agro-industrial raw materials' (World Bank 1994a: 15).

Post-independence phase

The phase following independence in 1980 is characterised by control of state agricultural support structures being taken out of commercial farmers' hands and expanded to include the former Tribal Trust Lands, renamed the communal areas. Policy was to provide services to communal farmers to raise communal production and incomes. The thrust of marketing policy was to build new depots and collection points in communal areas. 'The inherited array of market control mechanisms was extended to include fixed or floor prices for most crops, compulsory delivery of controlled crops to marketing boards, controlled consumer prices for major staples, and restrictions on imports and exports' (World Bank 1994a: 15) The expansion of Grain Marketing Board depots into the communal areas was accompanied by subsidised credit and input provision. The few smallholders in better served communal areas were the beneficiaries of the post-independence policy shift and responded with increased maize and small grains output, resulting in record levels of stockpiling by GMB. In effect the maize production revolution of the commercial areas in the 1950s and 1960s was extended to the fertile areas of the communal lands, using the state-owned, single channel marketing structures. The difference was that in the colonial period the costs of the state marketing agencies were borne partly by consumer levies and lower prices to smallholder farmers, whereas after independence the Treasury funded the costs of expanded state structures, with donor assistance (Jayne and Jones 1997: 1512).

A turning point in the post-independence period is the mid-1980s, by which point post-independence policies of expanding credit, extension and marketing services, and guaranteeing prices, had resulted in increased output of cotton and surpluses of maize and small grains in the good rainfall years of 1985 and 1986. These were sold at a loss by Grain Marketing Board, which had guaranteed prices to farmers. With the budget deficit growing, the remaining part of the post-independence period saw reductions in incentives (lower guaranteed prices, reduced buying by state agencies), and in real levels of operational budgets for extension and marketing

Transition to open markets

A third phase of policy begins in the late 1980s with the first moves towards open agricultural markets. The driving forces for change are the increasing costs of the marketing system in relation to Treasury resources to pay for it, and the worldwide ideological shift towards open markets, producing pressure – particularly from donors – to end state monopolies. In the late 1980s and early 1990s there was deregulation of domestic maize marketing, in the form of unrestricted movement of maize between rural areas and cities, ending of the Grain Marketing Board's monopoly on buying and selling maize, privatisation of some marketing boards (cotton and dairy), and corporatisation of others (grain and meat). By the late 1990s the task was begun of reorganising the Ministry of Agriculture itself to concentrate on its core functions. But this process has been overtaken by the political disruption following government-sponsored invasions of commercial farms, further economic decline, and suspension of donor aid to the process.

In sum, recent history is transforming the relation between the state and the agricultural sector:

1 Agricultural policy since 1980 has been dedicated to assisting the communal area farmer. It succeeded at first through expansion of marketing and extension services, particularly through subsidised inputs of fertiliser and pesticide. A weakness in its strategy was to

encourage increased food crop production, surpluses of which were disposed of at a loss. This contributed to the growing bad debts of the Agricultural Finance Corporation (AFC) whose lending had been switched towards communal farmers after independence. Increases in production did not continue and the state effort became too costly. Furthermore, the marketing policy – marketing agency monopolies and guaranteed producer prices – served the interests of large farmers and urban consumers more than communal farmers (World Bank 1996: 97, Collier *et al.* 1995: 182).

2 The state agricultural apparatus had been built up to assist commercial farmers. After 1980 the continuation of price guarantees and state monopolies still left large farmers and millers the main beneficiaries. But as regards agricultural services, after 1980, and especially since liberalisation of output and input markets, commercial agriculture began to rely on private agricultural services for technology, inputs and marketing, except in the case of livestock disease control (particularly veterinary fences), some services operating on a cost recovery basis to the commercial sector (milk quality control, veterinary services for wildlife, Agritex farm planning services), or in selling to GMB when guaranteed prices exceed market prices (e.g. 1993).

3 As a result commercial farming has needed less state agricultural services while communal farmers by the mid-1990s were no longer being as well served by them as they were in the post-independence phase, particularly in the less favoured natural regions.

4 This situation is not unique to Zimbabwe. It has occurred in high- and low-income countries as income and product markets develop and commercial agriculture is dominated by large farming companies. The imperative it brings is restructuring of state agricultural services towards assisting market development, particularly in communal and resettlement areas. There is general acknowledgement in Zimbabwe that communal agriculture must become commercial rather than subsistence-oriented, and that the long-term objective is an integrated farming sector. State agricultural services have an important role to play in the process. But to do so they must overcome obstacles of resourcing and management, as Figure 6.1 indicates.

Performance of arrangements

Zimbabwe's public agricultural services have been built up on a centralised, regulated, non-commercial basis. Chains of command are often long. Financial management is centralised. These characteristics have made its performance vulnerable to resource losses (staff and operating budgets) during the financial stringency of structural adjustment. But Zimbabwe's achievements in public agricultural services are threatened by declining financial and technical capacity and by the rigidity of provision structures which served a regulated agricultural industry well, but which are dysfunctional to market development. At the same time there are opportunities to benefit from market-oriented reform to achieve both cost savings and better service standards and coverage, especially through closer links in service provision to agricultural associations, private input suppliers, buyers, researchers and NGOs. Figure 6.1 summarises the above characteristics.

STRENGTHS	WEAKNESSES
• Great outreach • Ability to deliver if well-resourced • Powerful associational tradition • Excellent past record • Substantial assets • Many capable, committed employees	• Rigid • Partly immobilised by central funding cuts • Poor information flows • Weakened by loss of key staff • Unresponsive • Declining service standards • Outdated financial management
OPPORTUNITIES	THREATS
• A small amount of funding (for mobility and operating expenses) may yield substantial benefits • Strong interest in commercialisation in many units • Build on the success of the commercialisation of marketing parastatals • Service delivery links with agricultural associations, private input suppliers and buyers, NGOs • Organisational development in MOLA in mid-1990s provides good basis for team led change	• Continuing slow decline in funding basis • Failure to follow through current reform initiative

Figure 6.1 Swot analysis of agricultural services provision in Zimbabwe

To take a closer look at performance of government in the agricultural sector five aspects of performance are discussed:

- *Attaining objectives*: progress towards meeting stated objectives (effectiveness)
- *Focusing budget on services*: how much of the sector budget goes on service provision and how much is diverted to other purposes, measured by service provision/budget (allocative efficiency)
- *Providing the right combination of services*: decrease in services better provided by private sector, and increase in provision of public services needed for market development (allocative efficiency)[3]
- *Adaptiveness*: ability to reallocate resources intelligently to meet trends of change in demand for services (adaptive efficiency)
- *Positive overall management*: overall management of government services to the sector which increases morale, facilitates change and increases customer orientation (X efficiency)

Figure 6.2 overviews government performance on the aspects described above in each of the three periods in Zimbabwe's agricultural history. The approach is deliberately wide-sweeping, covering all aspects of policy and all services to agriculture. Performance is classified as 'High' or 'Low' in each case. Where performance is uncertain a '?' is entered. The classifications in all cases are subjective, reflecting the author's judgement. A discussion justifying the performance ratings follows Figure 6.2.

Discussion of performance aspects summarised in Figure 6.2

Attaining objectives (effectiveness): The mission of MOLA is 'to promote and sustain a viable agricultural sector based on the implementation of sound agricultural policies which optimise productivity through the provision of appropriate technical, administrative and advisory services' (MOLA Mission Statement 1997).

Pre-independence, the main target beneficiaries were commercial farmers and there was close political and operational integration between government services and the main target group. This made for generally high attainment of objectives. Other contributing factors include:

- public agricultural services organisations were smaller, with fewer but more experienced employees, thereby maintaining higher ratios of operational budgets to staff costs;[4]

Historical phase	Policy objective	Policy instruments	Stakeholder involvement in policy	Impact of wider public service environment	Performance in services to implement policy				
					Attaining objectives	Service provision/ budget	Providing the right combination of services	Adaptive efficiency	Positive management culture
Colonial (to 1980)	Promoting mainly white-owned commercial agriculture and processing	Closed markets, with state service monopolies	Producers' associations (mainly white commercial)	Positive, through maintaining real operating budgets and cash flow	High?	High?	High?	?	High?
Post-independence (1980–90)	Promoting communal farming and urban food price stability	Closed markets, with state service monopolies	Producers' associations, including communal	Initially positive, with increases in budget, then falling as inflation rises, real budgets fall and cash flow difficulties emerge	High at first in fertile areas, lower later	Falling, with increased subsidies to state monopolies	High initially, then falling as AFC debts mount and grains-oriented strategy produces loss-making surpluses	High initially as resources switched to communal areas. Falling with increased bureaucracy	Falling, with increased bureaucracy
Transition to open markets (1990s)	Promoting communal farming and urban foodprice stability	Opening markets, privatising parastatals	Rising stakeholder involvement	Budget and cash flow difficulties. Public Services Commission requires staff reduction	Low, as real budgets fall, food prices rise, and communal areas are poorly served by private services	Low, then rising with privatisation of marketing parastatals	Low, as real budgets fall and private services rise	Low but with potential to rise, as management reform begins	Rising with increased attention to improving management culture (MIPP programme). But lacking essential high-level political support

Figure 6.2 Overview of government's performance in agricultural policy and services in Zimbabwe

- the wider public service environment provided positive support to agricultural services, by maintaining real budget levels and managing cash flow so that operational units had funds and equipment to carry out their tasks.

The policy objective post-independence shifted to the more difficult one of developing communal farmers, by expanding services and shifting their focus within an otherwise little-changed structure. The initial rapid rise in communal output resulted from a number of factors, including agricultural credit, inputs and buying depots.[5] But relatively few communal farmers contributed the increased production[6] leaving the bulk of communal farmers with little benefit from the policy shift and extra resources devoted to communal areas. The cutback from the mid-1980s in expenditure on credit, guaranteed prices and inputs in response to surpluses of grains appears not to have been accompanied by cutbacks in service staff. Overall, despite some production success after 1980, there appears to be a fall-off in effectiveness in agricultural policy and services in the period as a whole. This is reflected in a slowdown in agricultural growth after the mid-1980s, leading to the conclusion that 'Zimbabwe's inward-looking agricultural strategy pursued over much of the post-independence period has yielded disappointing results, both in terms of growth and enhanced (household) food security' (World Bank 1994a: 18).

The later 1990s were characterised by recognition of this policy impasse, with market liberalising reforms, the emerging objective of an integrated, market-oriented agricultural sector, privatisation of parastatals, the beginnings of reorganisation of public agricultural services and of closer cooperation with private sector stakeholders.

Service focus of budget: The World Bank's Fiscal Management Review of Zimbabwe (1996) shows two distortions which reduce the proportion of public funds to agriculture spent on services.

The first is the large subsidy to the marketing boards and the Agricultural Finance Corporation (AFC) in the early 1990s, principally to the Grain Marketing Board (GMB) whose losses rose in 1993 to some 5 per cent of GDP as a result of fixing grain purchase prices above market prices and then exporting at a loss. Effectively this was a subsidy to grain farmers delivering to GMB and to international buyers. Privatisation of the dairy and cotton boards and lower GMB losses in the mid-1990s reduced these subsidies. But by the late 1990s, losses of the GMB and of the Cold Storage Commission (CSC), the meat processor, rose again.

The second distortion is the cutting of operational budgets within agricultural services, in order to protect salary levels. Real budget levels fell more than 30 per cent between 1990 and 1992, and stabilised thereafter. For 1995/6 only 5–8 per cent of the Department of Research and Specialist Services (DR&SS) budget went to research activities, with 68 per cent to salaries, compared to 50 per cent to salaries at independence (World Bank 1996: 102). In real terms the DR&SS budget was half its trend level between 1974 and 1990. The real budget for agricultural extension (Agritex) increased sharply after independence, but fell by the mid-1990s, with salaries taking about 70 per cent of the budget. Only the Department of Veterinary Services among the agricultural services agencies had a relatively low proportion of salaries in total costs (one-third). The strain on operations has been felt by all service agencies: field staff cannot travel easily; Agritex has altered its extension approach from individuals to groups, and has little coverage of the least fertile regions; research stations run out of funds early in the financial year; on-farm trials have been abandoned; and key research staff have left.

While subsidies to loss-making parastatals may now be declining there is little prospect of overall agricultural budgets increasing in real terms, meaning productivity gains will need to be found by reallocating expenditure.

Providing the right combination of services: The challenge facing government in a liberalised agricultural market environment is (1) to reduce provision of services which the private sector is increasingly able to provide more efficiently, and (2) to increase provision of essential public services needed for market development and stability – namely to provide food and inputs safety, to increase competition, to promote trade and innovation, to stabilise food markets in the face of shocks, e.g. drought, and to operate a policy process that is responsive and informed.

(1) Reduction of services which the private sector is increasingly able to provide: services related to marketing in which the state is reducing its involvement are most prominently the marketing boards which it has privatised (Dairy, Cotton) or is moving towards privatisation (Cold Storage Company and probably Grain

Marketing Board). These are also areas in which private firms have grown stronger. Other areas where the private sector is increasingly strong include seeds production and distribution, research, laboratory services, and veterinary curative services. But in these there has been less movement within government as yet to reduce its direct provision of services, or to involve private operators, or work with associations (e.g. on food safety, such as meat inspection, or phytosanitary control) – despite in some cases (e.g. research stations, seed development and regulation) the public services losing resources, market, scope and quality. There is movement towards user charging and other revenue generation to recover some cost. By the late 1990s departments were also increasingly being allowed to retain the costs recovered. In veterinary services a programme was underway to make use of private vets in remote areas to provide public services. Confronting the challenge of reallocating resources was the task the Ministry of Lands and Agriculture (MOLA) had set itself in the Agricultural Services and Management Programme (ASMP) with assistance from World Bank, EU and the UK.

(2) Increased provision of essential public services needed for market development and stability: meeting the increasing demand for public services which facilitate market development may require: skills and behaviour which are unfamiliar in a traditional bureaucracy; initiative in collaborating across former boundaries (ministerial, stakeholder); and taking the risk that the organisation which emerges may be smaller and different in structure and orientation from the former ministry, concerned more with impact, developing policy, managing contracts and working with stakeholder associations, than with providing direct services. Key areas in assisting market development are opening up trade while helping to ensure food price stability, competition, food safety, infrastructure development, innovation and assistance to target groups of farmers. Reducing transaction costs for the public in dealing with the ministry (e.g. for export and import licences) can be a further important contribution. The key measure taken in Zimbabwe to facilitate market development has been the opening of domestic agricultural trade[7] and – to a more limited extent (particularly in maize) – external trade. Relations with the public were also improved as was policy development and stakeholder involvement (notably through restructuring the

Agricultural Research Council and giving it some executive power). But despite progress, performance in market development has been held back by policy uncertainty over the opening up of external grain markets, particularly for the staple maize. It is argued in Chapter 7 below that opening external grain markets is important for market development and to provide food security at a lower social cost. Making a success of this will require close cooperation between government and private stakeholders, in an environment of improved information on crops, trade and stocks. The beginnings of such cooperation were evident in the setting up of the Grains Council, initiated in the private sector, but to which government has shown little commitment, tending to send only lower officials to meetings.

Ability to adapt (adaptive efficiency): The ability to adapt the use of public resources intelligently is vital for government which is committed to opening and developing markets. The reason is that market systems evolve and require change in the use of public resources to encourage further development of opportunities and maintain competition. Adaptive ability is the basis for good allocation of resources. It requires:

1 good information on which to base changes: on the impact of services provided (to enable services to be adjusted intelligently) and on trends and opportunities in markets (to enable interventions to maintain competition and stability);
2 good communication, both within government and between government and stakeholders;
3 senior managers committed to reallocating their resources in line with market needs;
4 a wider public service environment which supports change and provides the resources to assist change.

Figure 6.2 suggests a high adaptive ability immediately post-independence, as services were expanded and their focus switched from commercial to communal farming areas. Lower adaptive ability follows from the mid-1980s, as the post-independence initiative becomes less productive, resourcing becomes tighter, and the task of reforming the expanded bureaucracy becomes more challenging. Adaptive ability increases once more with the adoption of structural adjustment and market liberalisation in the 1990s, achieving privatisation of parastatals and turning to the task of adapting MOLA's own services, under the Agricultural Services and Management Programme (ASMP) supported by donors. But by 2000 the underlying tensions between government and the commercial agricultural sector had erupted into presidentially sponsored farm invasions, with donors suspending aid and rapid deterioration of the economy. Further, the permanent secretary of MOLA and his minister were both removed and charged with corruption. MOLA reform was an early casualty of these events.

Both adaptations in the recent past (post-independence and structural adjustment) resulted from crisis – the crisis resolved by political change at independence, and the budgetary crisis bringing about structural adjustment. The current crisis resulting from the invasions of commercial farms may bring further adaptation, as public services are geared up to support new small-scale farming settlements. In the long term, the hope must be that the liberalisation of markets and politics will create more flexibility in state structures so that adjustment to economic change will begin earlier and no longer require crisis (with all its attendant costs) to be achieved.

Positive management culture (X efficiency): the contribution of overall management of government services to the sector is to maintain morale, facilitate change and increase customer orientation. Together these increase 'X efficiency', so named because it is a relatively hidden factor determining the overall performance of an organisation. As with adaptive ability, the quality of overall management of agricultural services appears to have declined in the mid-1980s but to have improved in the late 1990s, despite the continuing resource constraints. The Management of Institutional Performance Programme (MIPP), working to develop more participation, communication and customer orientation within the MOLA headquarters, made a particular contribution:

> The basic assumption underlying MIPP is that the capacity to identify and implement solutions to the problems of the organisation is basically available within the organisation but needs unfreezing and remobilising ... At the beginning of the process, intensive consultations revealed a critical lack of clarity of roles, overlapping of functions, ineffective communication and lack of mutual trust. (Walker and Takavarasha 1997).

All persons interviewed in MOLA agreed that MIPP was of critical importance in improving communication, trust and the working environment in MOLA headquarters.

The key management challenge faced by MOLA was to focus on its core functions in the new environment of open and developing product and input markets. These were changes beyond the capability of an internal organisational development process. They required strong presidential support, and support from Ministry of Finance and the Public Service Commission – to allow greater retention of earnings (to encourage a commercial orientation), to enforce staff reduction targets and redundancy packages, and to ensure that key senior posts in MOLA would be filled by the most able staff. None of these was adequate to the task.

Conclusion: Zimbabwe

Changes in the role of government in agricultural markets in Zimbabwe have been driven by radical political change and, since 1980, by macroeconomic crisis resulting in liberalisation of domestic and external marketing – with the exception of the external maize trade, which remains controlled. Some success was achieved by the mid-1990s with privatisation of dairy and cotton marketing parastatals, though the grain and meat marketing parastals were not privatised successfully – owing to mismanagement and uncertain policy direction in the case of maize marketing, where confidence to privatise and liberalise fully was lacking.

Liberalisation has enabled the private sector and associations to begin supplying services formerly provided chiefly by the state (e.g. seeds, meat inspection, domestic grain trading), at the same time that funding for state agricultural services declined in real terms. The result was poorer quality and under-used public agricultural services. The state was faced with the task of reorganising to provide relevant and better services, geared to the needs of small farmers. By the late 1990s efforts in the Ministry of Lands and Agriculture to raise internal management standards provided a basis for further reform. But high level political support was lacking, particularly with regard to staffing changes.

The current political and economic crisis in Zimbabwe may precipitate further radical change in the role of government in agriculture. If it results in greater liberalisation combined with increased support to new settlement programmes for small farmers, a decentralised rural development effort rather than a ministry of agriculture may be appropriate, particularly since many of MOLA's traditional services would be carried out by the private sector.

7
Kenya

Peter Lewa[1]

Introduction

Through the colonial period and after independence in 1963 Kenya was the dominant economy of East Africa. However, the economy grew only intermittently in the 1980s and 1990s. In 1993, the government of Kenya introduced a programme of economic liberalisation and reform that included the removal of import licensing, price controls, and foreign exchange controls. Stop-go relationships with donors supporting structural adjustment, deterioration of infrastructure, civil instability, corruption and political uncertainty regarding the successor to President Moi hindering investor confidence, are factors variously put forward to explain Kenya's poor economic performance.

This chapter discusses the changing role of government in Kenya's maize economy, focusing on the reforms in maize trading and in the public marketing agency for grains the National Cereals and Produce Board (NCPB).

Government's involvement in the maize trade

Maize is the staple food for the majority of Kenyans and accounts for nearly half of the calories and usable protein available to the population. Maize is also a major source of income to many people. It is grown in both large- and small-scale farms. Small-scale farmers produce about 80 per cent of the country's total maize output. Large-scale maize producers are the main source of marketed maize. Maize production is strongly influenced by weather conditions. The timing, intensity, duration and the overall pattern of rainfall over any given year mainly determines maize yields. Over the years Kenya has experienced maize supply instabilities. It has been more than self-sufficient in maize in good crop years (e.g. 1982, 1983, 1987, 1989, 1990 and 1994), but in bad crop years (e.g.1981, 1984 and 1993) has imported large volumes. On trend Kenya appears to be moving towards a deficit in maize. The reasons for this (according to farmers interviewed, Lewa 1995 and 1998) include poor quality fertilisers, seed maize and falling prices after maize imports were liberalised in 1993. Cultivated area has stabilised at 1.4 million hectares and maize yield at 20 bags per hectare (MOALDM Reports, 1995).

The government's involvement in the maize marketing system goes back to before World War Two. A single channel state marketing network was put in place with controls on both domestic and external trade and guaranteed prices. The National Cereals and Produce Board (NCPB) was the organisation charged with running maize marketing. The Board was formed to regulate, control and manage the collection, storage, marketing, distribution and supply of maize and maize products; buy, store, sell, import and export maize in order to meet the needs of food security; and to advise the minister responsible for agriculture on the proper relation of maize to the needs of Kenya (Lewa, 1995; Act of Parliament No. 7 of 1985).

The government's policy objectives with regard to maize have tended to be food self-sufficiency, food security, price stability for maize producers and fair prices to consumers (Hesselmark and Lorenzil 1976; Schmidt 1979; Booker and Githongo 1988; Technosynthesis 1988) rather than efficiency in the marketing system.

But since independence in 1963, available evidence suggests that unacceptably high costs resulted from the government's domination of the maize marketing system. The monopoly position of NCPB led to operational inefficiency and high costs. The Board could not sustain the producer and consumer prices given the inadequate levels of exchequer support, it could not meet the national food security objectives due to poor management and high operational costs and also due to its capture by powerful political interest groups: grain farmers and those involved in the administration, transport and storage of grain for NCPB (Lewa, 1995; Lewa and Hubbard 1995). Markets were poorly integrated, private sector development was discouraged by the government monopoly, allegations of large-scale corruption and waste were rife, and NCPB ran up large deficits which were met by government. But by the 1980s carrying NCPB deficits proved impossible to a government facing structural adjustment.

Reform of maize marketing started in earnest from 1988 with the advent of the Cereals Sector Reform Programme (CSRP) sponsored by the European Union. CSRP was implemented in a stop-go manner over the next five years owing mainly to failure of the programme to build a reform culture with support at the presidential level, powerful opposition by the NCPB, sometimes conflicting signals from the various external donors, and owing to the varying maize gluts and shortages with their resultant financial and lobbying pressures on government (Lewa 1995; Lewa and Hubbard 1995; DAI and IDA 1989; NCPB papers 1996). The main achievements of the CSRP were liberalisation of domestic maize marketing in the late 1980s and liberalisation of maize imports in 1993. Despite an unstable beginning – liberalised maize imports were followed by a bumper maize crop in 1994, reducing prices – research (Lewa 1995) indicated that the benefits of private imports (readier availability of maize and no need for the government to finance imports) were quickly being appreciated.

Organisational arrangements and their performance

Since the growing operating deficits of NCPB were a chief reason for liberalisation, understanding why NCPB did not have the capacity to manage maize distribution successfully is central to understanding the need for change in marketing arrangements.

NCPB's powers

Maize marketing in Kenya has been characterised by a dominant role for the National Cereals and Produce Board (NCPB). The Board had wide-ranging powers in support of food security objectives, through a combination of monopoly status in maize trading, administrative controls over maize movement and prices, and control over private-sector maize trading activities. By its monopoly over maize trading, the Board was able to secure maize supplies from surplus areas and make these available to deficit areas through its depots or by licensing traders to deliver to those areas.

To ensure market access at reasonable and stable prices to producers, the NCPB invested heavily in developing a network of depots or intervention points, at which farmers could deliver their maize at pre-announced pan-territorial 'producer prices'. Consumers were assured of access to maize at controlled prices through officially determined ex-depot prices of maize, and controls over wholesale and retail price margins.

The NCPB depot network facilitated the accomplishment of market stabilisation, historically a key activity of the public sector. Without the public sector playing this role, it was argued that prices would be highly unstable as a result of Kenya's weather-induced variations in maize production and supply. Given that maize production in Kenya, until recent years,[2] moved between national deficit and surplus, the above argument found very strong support in the country.

The NCPB was charged with the responsibility of building up and maintaining food security stocks (strategic maize reserves) to promote national self-sufficiency in maize and to stabilise maize markets in times of inadequate maize production. Further motivation in this area was provided by the need to serve remote areas through the Board buying maize there.

The Board was also given the responsibility for maize imports and exports. Fears of local and national shortages of maize had always been cited as a reason for controls of imports and exports as well as controls over movement of maize within the country. The NCPB imposed further restrictions on the intake of maize by maize millers and on the distribution of their finished products.

Why the previous marketing arrangements failed

The single channel state monopoly for maize marketing proved unsustainable because government could not meet its own funding obligations for services provided by NCPB (mainly food security stocks' maintenance, grain imports for famine relief, and distribution costs for market stabilisation operations). By the mid-1990s a range of government departments had run up substantial debts with NCPB. Past shortfalls on government subventions for maize price subsidies (by the Office of the President) was the largest debt. Government ministries were the main debtors of the NCPB, mainly for famine relief activities carried out by NCPB. The ministries were involved in distributing maize for relief drawn from NCPB stores. Payments were not made immediately but promises to pay were given. In effect, government did not have the resources to fund the interventions that it undertook.

In spite of the unsustainability of government involvement in the maize marketing system, pressures against reform of the maize subsector, from powerful interest groups such as large-scale farmers and politicians with stakes in maize milling, continued even when the Cereals Sector Reform Programme (CSRP) was under implementation. It took pressure from the donors and gradual reform experience and support within government to achieve further reforms. Reform experience within government since cereals sector reforms began in 1988 helped confirm to those who initially opposed reforms that they had some positive effects and therefore deserved support instead of opposition (Lewa 1995; Lewa and Hubbard 1995).

Reforms

The complex reform process for maize marketing is described in Figure 7.1. It brings out the extent to which liberalisation was coincident with liberalisation of foreign exchange and introduction of multiparty politics.

External liberalisation since 1993

The liberalisation of the maize subsector in 1993 coincided with a period of short supply of maize in the domestic market due to poor weather in 1993/4 and general structural maize deficits in the country since the beginning of the 1990s (Lewa 1995).[3] However, the private sector responded very well, importing approximately 10 million bags, effectively averting food shortages (MOALDM Reports 1995). The private sector demonstrated its ability to import large volumes of maize using private finance. The distributive trade was able to supply all areas of the country with maize, with price differences between areas reflecting transfer costs,[4] suggesting that liberalisation had improved distribution.

By this time the government had started liberalising the foreign exchange system and so finances had started to flow in a free environment. Weather conditions in the 1994/5 crop season improved considerably resulting in a bumper harvest estimated at 34.4 million bags. This output added to the large imports by the private sector during 1994, leading to a situation of excess supplies in the domestic market. Farmers had no outlet for their maize due to export restrictions and also because most maize millers were already overstocked with imported maize (interview with maize millers 1995; NCPB records 1995). This situation led to serious dampening of producer prices to levels that could not even cover the cost of production.[5] The government took various measures to relieve the situation, including mandating the NCPB to purchase maize from farmers, import suspension, and maize exports through a tender system managed by the NCPB.

In November 1994, the GOK, in response to the farmers' plight mandated the NCPB to purchase maize from farmers to offload the excess amount that the private sector was not able to absorb.[6] This caused problems to the Board as it could not dispose of this crop due to the imported maize. As the bumper harvest of the 1994/5 season was not foreseen, there was no budgetary provision to deal with such a large surplus. The government had to reallocate funds within the existing budget to avoid exceeding the preset budgetary deficit ceiling under the macroeconomic framework.

By the time the NCPB intervened in the market there was already a glut mainly due to imported maize.[7] To deal with the problem of the

1986 –	study on NCPB reorganisation completed
1987 (17 Dec.) –	cabinet paper endorsing recommendations to reform the maize marketing system including NCPB reform launched
1988 (4 Jan.) –	Government requests the EU to help fund and implement the CSRP
1988 (25 Jan.) –	detailed CSRP implementation timetable launched
1988 (29 April) –	EU gives approval and support of the CSRP
1988 (June) –	announcement by the Minister of Supplies and Marketing that Cooperatives and Private Traders could buy maize directly from farmers
1988 (June) –	Maize millers granted licences through the NCPB to buy up to 20 per cent of their maize requirements from cooperatives and private traders directly
1988 (August) –	Decision authorising cooperatives and private traders to buy directly from farmers rescinded due to intense pressure from the NCPB, farmers and other vested interests
1988 (August) –	Kenya Grain Growers Cooperative Union (KGGCU) allowed to buy maize directly from farmers. The KGGCU was under the control of the large farm interests from the maize surplus districts
1991 –	due to mounting donor pressure, cooperatives and private traders allowed to ship maize without permit. Millers allowed to take up to 30 per cent of their maize requirements from private trade
1992 (Oct.) –	maize movement banned, licences given to maize millers and private traders revoked. NCPB's monopoly restored. Calls for multipartyism and donor pressure on the government to reform at their peak
1993 (Dec.) –	freedom to move maize and to import maize given to maize millers and other actors primarily due to lack of foreign exchange in government. Government technically insolvent
1994 (Aug.) –	government bans importation of maize due to adequate maize supplies from excessive imports, local supplies of maize and pressure from the NCPB. Decision rescinded soon thereafter due to intense pressure from donors, maize millers and maize traders
1995 (April) –	a 6–9 months' ban on maize imports imposed due to dumping of cheaply imported maize as per the government's announcement. Excessive maize imports during the last six months of 1994 had coincided with bumper long rains harvest of 1994/95

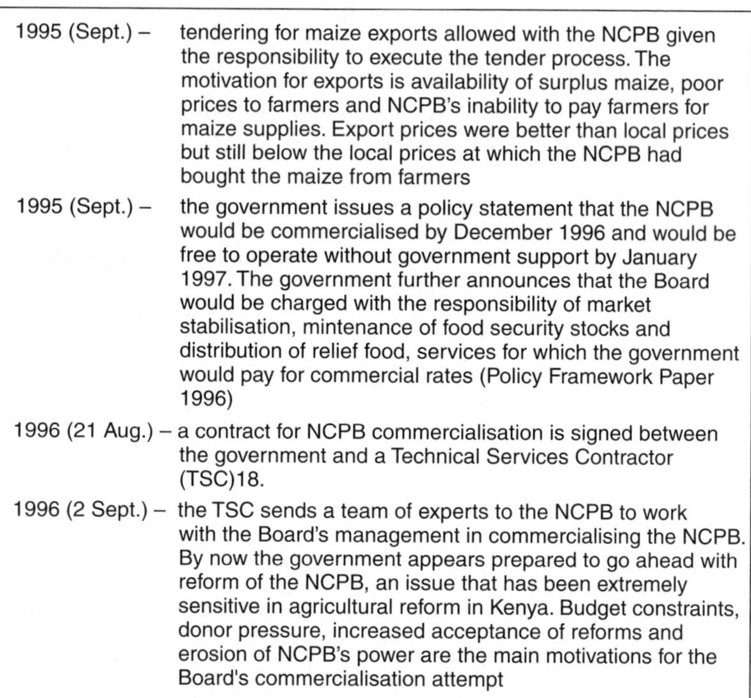

1995 (Sept.) – tendering for maize exports allowed with the NCPB given the responsibility to execute the tender process. The motivation for exports is availability of surplus maize, poor prices to farmers and NCPB's inability to pay farmers for maize supplies. Export prices were better than local prices but still below the local prices at which the NCPB had bought the maize from farmers

1995 (Sept.) – the government issues a policy statement that the NCPB would be commercialised by December 1996 and would be free to operate without government support by January 1997. The government further announces that the Board would be charged with the responsibility of market stabilisation, mintenance of food security stocks and distribution of relief food, services for which the government would pay for commercial rates (Policy Framework Paper 1996)

1996 (21 Aug.) – a contract for NCPB commercialisation is signed between the government and a Technical Services Contractor (TSC)18.

1996 (2 Sept.) – the TSC sends a team of experts to the NCPB to work with the Board's management in commercialising the NCPB. By now the government appears prepared to go ahead with reform of the NCPB, an issue that has been extremely sensitive in agricultural reform in Kenya. Budget constraints, donor pressure, increased acceptance of reforms and erosion of NCPB's power are the main motivations for the Board's commercialisation attempt

Figure 7.1   Chronology of events in maize marketing reform in Kenya

highly subsidised imports, the government imposed a six-months' ban on imports from 5 April 1995. This was also aimed at enabling the NCPB to offload the accumulated stocks from local sources and to streamline commercial and food aid imports which were distorting the domestic market (MOALDM Reports 1995). Initially this measure did not result in any significant sale of stocks by the Board because the market was holding stocks and maize imports from previous commitments were continuing to come into the country. The government then imposed an import levy on maize imports entering the country after the expiry of the ban – a move which was severely criticised by both donors and private traders, since it created uncertainty and was contrary to the intention of the reform.

Given the inability of the domestic market to absorb NCPB stocks, the government authorised the Board to tender for export up to two million bags of maize in May 1995. The tenders were floated in June 1995. Previously, during the period of controls, the Board used to look for markets for exports of maize and carried out the export programme itself. However, since liberalisation an important innovation has been adopted whereby exports are undertaken through a tender system.

The offers for the maize exports in May 1995 were generally low, with FOB prices ranging between Kshs. 403.40 and Kshs. 797.10 per 90 kg bag. Adding transport and handling costs of Kshs. 380 per 90 kg bag gave NCPB an average net realisable value of about Kshs. 290 per 90 kg bag. This was far below NCPB's ex-depot price of Kshs. 1280 (NCPB Reports 1995).

The transport costs problem

The long distance between Kenya's main port, Mombasa on the Indian Ocean, and the major maize-producing areas in the west of the country results in high freight charges for imports and exports of maize, a high bulk product. This gives increased protection to domestic maize producers in the main market, Nairobi. But it means that exporting maize is not commercially attractive and that the coastal areas of Kenya may find it cheaper to consume imported maize once imports were liberalised, which has happened, as it did in South Africa after maize market liberalisation (Bayley 2000). It also means that commercial maize imports and exports to and from the interior will only take place in response to sizeable shifts in prices – shifts large enough to overcome the transport costs – thereby reducing the role that external trade can play in stabilising domestic prices.[8] Transport costs are raised by the poor state of the Kenya–Mombasa road, poor wagon services offered by the Kenya railways, excessive delays at the port of Mombasa, corruption and too much documentation.[9]

Conclusion: Kenya

As Kenya moves into a liberalised maize market, changes to incentive structures, maize production, and roles of all actors including government are emerging. Greater price stability in maize will only be achieved as more efficient markets develop, encouraging storage and forward transactions to hedge the price risks resulting from Kenya's unstable maize production. This will require reducing transport and other sales costs, in which government must have a major role, particularly in infrastructure provision. Reducing the risks to people's welfare caused by maize production instability requires diversification of both production and consumption, as well as higher incomes for the poor. Liberalisation is encouraging farmers to switch out of maize into higher value crops (by reduced and less certain farm-gate prices where NCPB is no longer buying at guaranteed prices) and the public to include other grains (rice, wheat) in their diet.

Part III

Key Issues in Developing Agricultural Trade

8
Can Food Supplies be Entrusted to the Market?

The central question for governments in liberalising the grain trade is whether food supplies can be entrusted to the market. Previous chapters suggest the increased orientation by countries worldwide to making markets the basis of their staple food systems. The key practical issues are, first, how the process of deregulating the grain trade internally and externally can be managed politically, without disrupting food supplies or markets for farmers; and second, how the staple food trade can be enabled to work efficiently, cover the whole country and be accessed by the whole population.

This chapter first identifies features of the process of agricultural market liberalisation, on the basis of the experience of the case study countries and others. The tasks that governments must successfully carry out if liberalised agricultural markets are to provide national food security and stimulate agriculture, are then discussed. Finally, the capabilities states need to carry out such tasks are overviewed.

Features of the process of agricultural market liberalisation

The process and tasks of agricultural market liberalisation differ markedly between countries because of the varying extent of their intervention in marketing, development of private trade, political strength of their farming lobby and the strength of public finances, among other factors. The basic task is reducing barriers to private domestic and external trade. Other tasks variously include abolishing public marketing agencies or greatly reducing their role, and reducing consumer and producer subsidies.

The most challenging transition is where food prices, farm-gate prices and inputs were subsidised, all trade was monopolised by a state organisation, and grain production is by peasant farmers, many of them in remote areas, in a climate prone to crop failure where relief was provided by state food rationing. Under these circumstances government is faced with a major reorientation towards private trade and substantial rural development and safety-net provision.

In Part II above the case country experiences with liberalising agricultural markets were discussed. The main observations were as follows:

India: The food policy priority continues to be national food self-sufficiency. There are substantial subsidies to farmers and large-scale buying, storing and selling by state agencies, and much interference with domestic trade in foodgrains. Though effective in times of emergencies (e.g. drought), inefficiencies and lack of targeting mean the public distribution system is often ineffective in getting food to the poor in normal times and imposes heavy fiscal costs. The system also distorts farm production towards food grains, has restricted development of food trade (particularly storage) and infrastructure, and causes grain to be marketed in seasonal peaks, with consequent destabilising effects on prices which then must be stabilised by state sales and purchases. External food trade liberalisation has proceeded further than internal. Private rice exports from India to traders in Bangladesh enabled rice prices to be stabilised in Bangladesh following poor harvest and floods in 1997–8 (Dorosh 2001).

Sri Lanka: Liberalisation focused on diminishing the role of the Paddy Marketing Board. The private sector has played an increasingly important role in rice importation, while innovative means of providing food security (bonded rice imports) were developed and efficiency of the marketing system was maintained. Wheat importation and milling were made the responsibility of a licensed monopoly. The broad trend of reform since 1980 has favoured liberalisation and growth of the wheat market against the rice market, evidenced by falling real wheat flour prices and consequent downward pressure on real rice prices. Conflicts in food policy have remained, with the incoming government in the mid-1990s trying to support both rice farmers and poorer urban consumers. But its renewed buying of rice via the Paddy Marketing Board, to support farm-gate prices, was not financially sustainable and ended in 1996.

Ghana: The diversity of Ghana's staple diet (root crops, rice, maize, wheat), with private trade playing the dominant role,[1] and concurrent increases in food production, mean that liberalisation has not increased food security risks. Realising further benefits of liberalisation is hindered by low capacity in government – in policy formulation, taxation, public expenditure management and rural infrastructure provision.

Zimbabwe: Liberalisation of internal agricultural markets in the early 1990s enabled the private sector and farmers' associations to begin supplying services formerly provided chiefly by the state (e.g. seeds, meat inspection, domestic grain trading), at the same time that funding for state agricultural services declined in real terms. Failure to adjust public agricultural services left their quality and relevance in decline. By the late 1990s efforts in the Ministry of Lands and Agriculture to raise internal management standards provided a basis for reform. But high-level political support was lacking, particularly with regard to staffing changes. Reduction of state maize purchasing left smaller maize producers with much-reduced market access, which the private sector has not filled owing to poor infrastructure and intermittent, scattered surpluses in small farmer areas. External maize trading has not been liberalised. Distrust by government of the predominantly white private sector, culminating in current confiscation of white-owned farms, leaves commercial agriculture with an uncertain future.

Kenya: Liberalisation of domestic marketing and of maize imports is encouraging farmers to switch out of maize into higher value crops (by reduced and less certain farm-gate prices where NCPB is no longer buying at guaranteed prices) and the public to include other grains (rice, wheat) in their diet. Maize consumption in the coastal zone has been increasingly supplied by imports through Mombasa, as a result of the high internal transport costs for maize. But the poor in remote areas, farming maize on a semi-subsistence basis, have not automatically benefited from liberalisation, finding it less easy to sell any maize surplus or to buy maize cheaply in times of shortage.

These experiences fit into the broader picture emerging from the literature:

Underlying staple food market liberalisation tensions in developing countries is the policy conflict between providing cheap food to the urban poor and supporting grain farmers, among whom are many of the rural poor. In wealthier countries, such as the EU and US, there have been the resources to subsidise both. In poorer countries, attempts to serve both constituencies have resulted in financial crises, increased donor involvement and stop-start liberalisation. Liberalisation has provided budgetary relief (by reducing costs of subsidies to marketing agencies) and lowered food costs to many urban dwellers (particularly those near ports), but faces grain farmers with lower prices.

Sectional interests and uncertainty drive the resistance to liberalisation. Potential beneficiaries of agricultural market liberalisation tend to be taxpayers, urban consumers (though they may lose in the short term when consumer subsidies are removed), and many private traders. The most direct losers have been those for whom trade restrictions provide jobs and protection from competition – particularly grain farmers and employees of public marketing agencies. They have often been a more concentrated political lobby (see examples in Jayne and Jones 1997), reinforced by anxiety among politicians that liberalisation would cause greater food price instability and deprive the rural poor of access to grain sales and purchases at subsidised prices. As a result, the liberalisation process in most developing countries has been halting and protracted.

Public finance crises have accelerated liberalisation. With few exceptions, e.g. South Africa (Bayley 2000), liberalisation of agricultural trade has been precipitated by crises in the public finances. In some cases the crisis resulted directly from massive, unsustainable debts of the public grain marketing agency (Zimbabwe, Kenya, Zambia). In others (e.g. Sri Lanka, India, Ghana) a generalised financial crisis was the source. In all cases the crises resulted in structural adjustment agreements with donors and multilateral agencies (IMF, World Bank), which tilted the policy balance in favour of liberalisation.

Liberalisation is an uneven and ad hoc process in many countries. Political struggle among sectional interests results in some markets being liberalised sooner and further than others. Domestic grain trade was the first to be liberalised in Kenya and Zimbabwe, while resistance to liberalising grain exports continues. In India, external grain trade has been liberalised faster than domestic trade. In Zambia, agricultural input markets were liberalised later than the grain trade (Pletcher 2000) in contrast to Zimbabwe. The Kenyan liberalisation experience demonstrates how long and ad hoc the process can be, and the need for robust political leadership to make it succeed (Lewa and Hubbard 1995). In the 1980s and early 1990s, the lack of examples of successful liberalisation probably made the process more tentative for policy-makers.

Countries with low state capacity have experienced the greatest difficulties with managing the process and impact of liberalising agricultural markets, yet have often liberalised most rapidly. Reform has been heavily donor-driven in countries such as Zambia, Tanzania and Malawi. Immediate gains of liberalisation are for the public finances, while the process of liberalisation risked disruption of markets (as occurred in Zambia in 1993 when the National Agricultural Marketing Board ceased trading, with urban food riots in 1992 following removal of food subsidies; Seshamani 1998). Many poorer countries have large numbers of semi-subsistence grain producers in remote areas, who lose from the withdrawal of state grain purchasing and sales where private trade is weak, and where the state is unable to create conditions in rural areas for private trade to flourish. Malawi (Chilowa 1998) and Malagasay (Barrett 1998) are cases in point.

More successful liberalisers of agricultural markets tend to have some or all of the following characteristics: greater state capacity, more developed agricultural and financial markets, less crisis-driven relations with donors. Some attempt more creative approaches to food security. South Africa is a radical recent liberaliser, with good results to date. Helped by the transition in political power to black, mainly urban interests, away from white farming interests, South Africa in the mid-1990s abolished internal and external grain and meat marketing controls. At the same time the parastatal control boards which controlled previous arrangements, were abolished. The South African Futures Exchange (SAFEX) was formed and in 1996 began trading futures contracts in white and yellow maize, which allows price risk management by millers and traders, more secure prediction of produce value by farmers, and buffers the maize meal market from producer price instability (Bayley 2000: xv). Factors behind the success include: established commercial agriculture, finance and trade; good infrastructure; storage and transport facilities already in private hands (the control boards worked through contracted private agents); plus the political will and administrative capacity in the state to manage the reform.

Sri Lanka managed liberalisation of internal and external trade in food grains in the 1980s while maintaining a well-functioning marketing system (see Chapter 4 above). To ensure food security supplies the innovation was adopted of licensing rice importers to maintain minimum rice stocks in bonded warehouses. Management of supply chain risk was thereby shifted by regulation on to the private trade. South Africa achieved this by developing forward markets.

Bangladesh stands out as a case of successful close collaboration between government, donors and NGOs. Bangladesh has increased rice production, thereby reducing its trade gap in rice, at the same time as liberalising domestic and external trade in food grains and farm inputs, and switching to more targeted food subsidies. Counterpart funds from selling food aid played an important role in facilitating policy change

(Goletti 1994; Ahmed, Haggblade and Chowdhury 2000). Liberalised external trade enabled commercial rice imports from India to balance rice prices following Bangladesh's poor rice harvest of 1997 and floods of 1998 (Dorosh 2001).

Food aid has been important in donor support to liberalisation programmes, but risks changing food habits towards imported wheat. In Bangladesh food aid from donors helped drive through the liberalisation reforms in the 1980s and early 1990s (Ahmed, Haggblade and Chowdhury 2000), with counterpart funds from sales of food aid providing an important source of state funding for targeted welfare programmes. Sri Lanka had a similar experience, where there has been concern (Smith and Ellis 1997: 25) that cheap food aid in wheat was reducing consumer preferences for domestically grown rice, as discussed above.

Food market development: tasks for the state

For liberalised agricultural markets to deliver food supplies well and to stimulate agriculture, the state has continuing and new tasks in assisting food market development. These are discussed as tasks to encourage deepening of staple food markets to provide reliable food supplies, and widening of staple food markets to enable more people from all areas of the country to buy and sell a greater variety of food. Well-functioning food markets strengthen the supply side of food security. Ensuring that all have the purchasing power to buy staple food remains a critical function of the state.

Deepening food markets to manage supply risk

Developed markets serve food security by operating a marketing chain from farmer to consumer, based on contracts for purchase and delivery. They have 'depth' in that they involve layers of forward contracts – from farmer to wholesaler to processor to retailer, thereby linking supply from farmers to consumers on a continuous basis. The risks inherent in forward contracts (e.g. crop failures, changes in market prices) are reduced through futures and options trading in the commodity,[2] particularly by processors who require guaranteed delivery of supplies in order to meet contracts with retailers. In this way the market anticipates changes and participants insure themselves against losses. Grain storage in anticipation of price rises plays a key role, particularly in helping to reduce inter-seasonal price differences, as do functioning finance and foreign exchange markets. Clear property rights in traded grains are needed, such as tradable warehouse receipts (Coulter and Onumah 2001; Larson, Varangis and Yabuki 1998), as well as low transport costs and a minimum of arbitrary intervention by the state in domestic or external trade, so that political risks are low. Well-functioning processing and retail sectors are needed, which in turn rely upon steady consumer demand. Up-to-date information regarding expected crop levels and prices is essential so that market participants can adjust their risk strategies by buying and selling futures and options.

Commodity exchanges have played a key role in grain market development, serving to set prices, provide and enforce standard contracts (through membership of the exchange), arbitrate disputes, develop forward, futures and options trading, and disseminate prices and forecasts. Commodity exchanges rely on volume for success: the higher the volume the greater the anonymity of deals, the better the deals that can be made, and the less chance of getting a better price by direct trading between a single buyer and seller, cutting out the exchange. Exchanges benefit buyers and sellers generally by providing price leadership. With instant global communication commodity exchanges now serve wider markets. A country does not need to have its own commodity exchange to benefit from the services exchanges provide to trade. For small farmers to benefit from commodity exchanges an intermediary organisation is required which aggregates their risk. For example, ASERCA (a cooperative marketing association in Mexico) hedges cotton prices to producers by using their fee to buy a 'put option' on the New York cotton futures exchange, which gives it the right to sell a given quantity at a given price on a specified date (Larson, Varangis and Yabuki 1998: 21).

Among the case study countries, Zimbabwe and Kenya in the 1990s both opened commodities exchanges on which staple foods were traded. By contrast, India, with long-established commodity exchanges trading futures contracts for a variety of cash crops, excludes food grains. It is currently considering extending futures trading to food grains, recognising that for this to succeed dismantling of state interventions would be required. Sri Lanka, too, is at the planning stage for futures trading in food staples.

The role of the state in assisting food market development consists both of direct encouragement to the food trade sector (clear and consistent food trade policy, facilitating key investments, ensuring competition, working with food industry associations on food trade problems) and of ensuring the functioning of the wider markets and services on which the food trade relies (finance, foreign exchange, property rights, ports, transport and communications, customs).

With the possible exception of Sri Lanka, the performance of the state in the case study countries in encouraging market development in staple foods has been poor. This contrasts with often substantial state efforts in the same countries to promote agriculture. Distrust of the private trade, particularly in staple food marketing, has been slow to change. In Kenya, and particularly in Zimbabwe, the distrust has been deepened by a predominance of ethnic minorities in trade. In India, continuing administrative obstacles to domestic private trade in staple foods have restricted the development of storage and forward markets.[3]

Improving the wider public services on which the private food trade relies has been hampered by fiscal crisis in governments. High interest rates have been common where governments have financed their fiscal deficits by borrowing domestically, discouraging banks from developing new lending outlets, such as trade finance. Capital spending by governments has been constrained by budget deficits, restricting the infrastructure investments on which trade growth depends.

Incomplete commercialisation of public marketing agencies restricts private trade development by adding uncertainty regarding how the agencies will be used by government,[4] and keeping much storage capacity in public hands.

Two examples of how initiatives to deepen staple food markets have been hindered by incomplete commercialisation of public marketing agencies are the following:

Inventory credit in Ghana

In Ghana a scheme was begun in the early 1990s to improve finance for domestic trade in maize by using stored maize as collateral. This is the Inventory Credit Scheme (ICS) started by the state-owned Agricultural Development Bank with NGO and donor assistance. With most of the traders being small-scale and inexperienced in managing credit,[5] access to institutional credit has often been limited by lack of suitable collateral. Pilot implementation of the scheme appeared quite successful (Coulter and Asante 1993) with a multinational bank (Barclays) voluntarily getting involved in the 1995/6 programme because of satisfaction with the results. It also seemed to have encouraged literate entrants with technical expertise into the maize trade (Coulter and Shepherd 1995).

Further expansion, however, depended on raising enthusiasm of banks for new forms of lending, which had been dampened by easy profits to be made from lending at high interest rates to government to finance its deficit. Expansion also depended on resolving problems with inventory management. The Ghana Food Distribution Corporation (GFDC), which was the only agency with significant storage and grain drying facilities in Ghana in the early 1990s (World Bank 1992), had acted as inventory manager for the scheme, guaranteeing the security and quality of stocks held as collateral (Coulter and Asante 1993). But conflict of interests in its role as inventory manager and commercial grain trader seemed to be undermining its effectiveness in fulfilling this role (ibid.). Despite this hindrance, the inventory credit initiative has been able to expand and make a promising contribution to finance for farmers and traders, with replication in Mali and South Africa.[6]

An agricultural commodity exchange in Zimbabwe

The Zimbabwe Agricultural Commodity Exchange (ZIMACE) was established by the Commercial Farmers Union and a local stockbroking company (Edwards and Co) in 1994. It was the first agricultural commodity exchange in Africa, pre-dating the South African Futures Exchange (SAFEX) which began trading commodities in 1995, the Kenya Agricultural Commodities Exchange (KACE) and the Zambian exchange. It emerged to fulfil the need, arising from the liberalisation of the domestic market in Zimbabwe's Economic and Structural Adjustment Programme (ESAP), for an orderly, transparent market, whose prices would provide a guide for other sellers and buyers.

New exchanges face the challenge of getting known and trusted among farmers and traders and settling procedures down, so that a critical level of volume to sustain the exchange can be attained. Early growth resulted in ZIMACE handling about one-third of all marketed grains. But many large transactions were still taking place through direct deals (e.g. between farmers and brokers for millers, saving farmers the costs of employing their own brokers). This practice was outlawed by the ZIMACE board in the late 1990s, preventing brokers with a seat on ZIMACE from buying directly from farmers. This was a challenge to brokers to support ZIMACE or abandon it, which the exchange appeared to be winning.

A further challenge for an exchange is to expand into external markets. This requires a functioning foreign exchange market and active external trade. ZIMACE was founded on the hope that Zimbabwe's position as a major regional grain producer, situated centrally in southern Africa, surrounded by grain deficit countries, could make it a crossroads in the southern African grain trade – a favoured place for storage and re-export, for forward and spot contracting, bringing together a variety of commodity buyers and sellers. To become such a crossroads requires not only the locational advantage that Zimbabwe has at the centre of the region but also superior convenience in communications, transport, efficiency of customs clearance and commercial storage. The food security advantages of becoming a crossroads in the regional grain trade would be substantial: in particular it would increase stocks of marketed grain in the country, with the trade taking on the costs of stocking. This would create the possibility for managing any physical strategic grain reserve within the trade on a bonded basis, as in Sri Lanka,[7] with low public costs.

For ZIMACE to prosper required government policies favouring market deepening in agricultural commodities. The structural adjustment reforms of the early 1990s were a favourable step, liberalising domestic trade and giving the Grain Marketing Board (GMB) commercial independence, with government paying for any non-commercial activities which it required GMB to undertake. A continuing GMB monopoly of maize exports and imports – including evidence that at times in the mid-1990s GMB used this monopoly to raise local prices to improve its own trading results – was seen as a problem which would be resolved as government gained increasing confidence in the ability of the market to stabilise prices. By the late 1990s, government had eased restrictions on private imports of maize, as Kenya had done in 1993, and GMB had taken a trading seat at ZIMACE and begun to lease its extensive grain stores to the private sector. Such developments compensated for the lack of direct encouragement from government to private sector growth (e.g. ZIMACE struggled initially to get adequate phone connections).

However, any prospects for ZIMACE success have been put on hold as the Zimbabwe government in July 2001 reimposed a GMB monopoly on all maize trading (domestic and external) as a result of maize shortages. ZIMACE ceased trading.[8] The shortages and escalating maize prices were caused by much-reduced maize production following commercial farm invasions and drought. A market development process which had shown positive achievements in raising food security (see Jayne, Rubey *et al.* 2001) waits to be restarted.

By contrast, with the strong backing of the South African government, SAFEX's agricultural commodity trading has expanded rapidly, developing an options market in 1998.[9] Government provided a clear and stable policy framework with the passage of the Marketing of Agricultural Products Act, 1996, which abolished the commodity control boards. This was reinforced by the relaxation of exchange controls, liberalisation of agricultural exports, and consistency in pursuing liberalised market solutions. During the mid-1990s transition from a controlled market in the mid-1990s, maize producers, cooperatives and processors had lobbied for the retention of established mechanisms for managing their risks, namely a floor price and export subsidies. Government resisted these lobbying efforts and market participants turned to SAFEX instead – thereby ensuring its expansion – since both farmers and processors need to manage their risks (Bayley 2000: 94).

SAFEX provides potential benefits for other countries in the region, as a means of helping to manage grain market risks for grain farmers and processors through external trade. Since currencies of countries in the region are substantially pegged to the South African Rand, foreign exchange risk is reduced. As regional trade increases, contracts (either SAFEX or a local exchange linked to SAFEX) based on silos in or nearer to countries in the region become feasible (Bayley 2000: 94). These would more closely reflect local supply risks and reduce costs of delivery in contract fulfilment. Taking full advantage of SAFEX will require countries in the region to build an active and competitive external trade in grains, and minimise intervention so that trade confidence is maintained.

In sum, market deepening is essential if private trade is to provide food efficiently to the major food deficit areas, particularly cities. This requires high volumes, delivered reliably, with low variation in price and quality. The discussion above demonstrates how trade liberalisation assists market deepening. But trade liberalisation does not necessarily benefit rural consumers and smaller grain farmers in the short term. Market deepening may favour larger producers who are better able to manage contracts, so may narrow the market by discouraging smaller producers. Grain producers and consumers in remote, underdeveloped areas can be the most direct losers, as public marketing agencies close their remote area depots, and cease to buy and sell grain at subsidised prices. In Zimbabwe, after liberalisation of internal grain trade in the early 1990s, the private grain trade developed little in remoter areas formerly served by GMB depots. Subsequently many farmers were unaware of prices or when a buyer would be available (Chikandi 1999). A pilot market information system for peasant farmers distributed current ZIMACE prices (see Chapter 12 below). But small, sporadic surpluses from scattered farmers do not easily build a regular private trade. The result has been complaints from politicians that private trade exploits such areas by offering prices well below those in more accessible areas. This raises the problem of how to make rural food markets function well. It is discussed below as a problem of widening markets.

Widening staple food markets

The widening of staple food markets involves bringing more producers and consumers across the country into the market, and the selling of a wider variety of produce. The demand side of widening staple food markets requires higher consumer incomes and more diversified tastes. The supply side involves stimulating competitive food production and trade. There is no formula for widening food markets, since wide-ranging rural development is required. The literature exploring development of the rural non-farm economy suggests key factors are growth of resource-based incomes (agriculture, forestry, mining, tourism), and growth of rural towns, population and infrastructure (Haggblade *et al.* 1989; Reardon 1997; Reardon *et al.* 2001). In Zimbabwe, Kenya and Ghana, there appeared to be scant progress with rural development in the liberalisation phase (late 1980s–mid-1990s) and governments provided little leadership for the task, beyond a fashion for decentralisation which has yet to make any substantial impact.[10] In India and Sri Lanka studies suggest a degree of integration between staple food markets in different regional centres (Palaskas and Harriss-White 1993; Smith and Ellis 1997), while rising Human Development Index scores, steady per capita income growth, and little fall in the rural proportion of the population, indicate a more buoyant rural economy (World Bank 2000; UNDP 2001). In Sri Lanka's case absolute numbers in poverty have fallen in the last ten years (UNDP 2000: 109).

Liberalisation of trade encourages diversification of diets where trade controls had previously reduced access to a food which is in demand. But the main diversification favours wheat consumption, often imported – as in the maize staple economies of southern and east Africa, and Sri Lanka. Ghana stands out among the case study countries for the greater diversity in its staple diets (rice, maize, wheat and a variety of root crops), and the related fact that public marketing agencies never dominated food marketing. Enabling local agriculture to benefit from liberalisation and cheaper transport (e.g. airfreight) by diversifying into non-traditional crops, either for exports or local markets, is critical. Fresh flowers, fruit and vegetables are examples of new long-distance trade which has become established in recent years from low- to high-income countries as a result of reduced air freight costs (Jaffee and Morton 1995). The expansion of unregulated raw milk sales by cooperatives in Kenya following the ending of the urban milk market monopoly of Kenya Cooperative Creameries is an example of local market expansion (Owango *et al.* 1998).

Often remote areas are potentially more accessible to the transport and communications of neighbouring countries. But development of regional cross-border infrastructure is low, with few countries prepared to commit resources to developing other than the main regional arteries of trade and transport, or to encourage an easy flow of people across multiple border points for trade or employment.

Ensuring all can buy staple food

Since agricultural markets are the focus of the present work, social safety nets to ensure that all can buy staple food are discussed only briefly.[11] Shifting to market-based food security allows poverty relief to be changed from state purchase and distribution of food to income-based relief, in the form of cash or vouchers (e.g. food stamps) to enable purchase of food. The cost of purchasing, storing and transporting food supplies is thereby transferred to the private food trade, while government takes up the task of financing and distributing income-based relief (e.g. through labour-intensive public works). In both food-based and income-based relief there is a fiscal cost to be met.[12] Targeting relief to the poor is the main strategy for reducing the fiscal cost. For example, Sri Lanka as part of its structural adjustment programme in 1980 moved 50 per cent of the population off the rice ration scheme and subsequently built up a food stamp programme. Some Indian states – most prominently Maharashtra via its Guaranteed Employment programme in the 1980s (Hubbard 1988) – give cash payments not food in their famine relief employment programmes, which are self-targeting to the poor since work is hard and pay is low. Botswana, and Zimbabwe to a lesser extent, have set up similar income-based relief schemes during droughts.

In addition to such income safety nets to help deal with shocks, several developing countries (e.g. Chile, South Africa, Namibia – see World Bank 2000: 154) have begun social pensions programmes to reduce chronic poverty. These require considerable financial and administrative capacity in the state and are best provided alongside private pensions, so that a multi-pillar system of social protection is built. A well-developed financial sector provides the basis for private pension expansion.

Conclusion

There is now much evidence (e.g. Kenya and South Africa since the mid-1990s) that a developed food trade greatly strengthens food availability, provided that incomes of the poor are maintained so there is a steady market for food.

Liberalisation is an uneven and ad hoc process in many countries, with sectional interests and uncertainty resisting change and public finance crises accelerating change. Countries with low state capacity (financial, managerial) have experienced the greatest difficulties with managing the process and impact of liberalising agricultural markets, yet have often liberalised most rapidly (e.g. Zambia, Malawi) partly owing to donor pressure. More successful liberalisers of agricultural markets tend to have greater state capacity, more developed agricultural and financial markets, less crisis-driven relations with donors, while some attempt more innovative approaches to food security (Sri Lanka, South Africa).

Although embarked on market liberalisation, governments in the case study countries have yet to proceed confidently with market deepening and widening tasks required to maximise the contribution of private food trade to food availability. Similarly, the task of developing income-based safety nets is at an early stage.

9
Can Public Marketing Agencies be Reformed?

Can public marketing agencies be reformed and play a productive role in developing agricultural markets, either by becoming competitive private firms or by providing public services beneficial to trade? Or is dissolution the only option when policy is to encourage market development? This chapter focuses on reform of public marketing agencies in liberalisation of agricultural marketing, principally in Zimbabwe, Sri Lanka, Ghana, India and Kenya, in the light of theory regarding commercialisation of public agencies.

Reforming public marketing agencies in the case study countries

Public marketing agencies carry out functions which are classified as mainly private in the theory of public and private goods[1]: buying, transporting, storing and selling produce – services which private firms undertake if allowed to, and if there is enough profit in doing so. Therefore efforts to reform public marketing agencies have focused on developing their commercial potential. In general, commercialisation of public agencies has taken place in three stages, as outlined in Figure 1.2 in Chapter 1. First, management is strengthened (restaffing, training, computerising accounts, business plan with performance targets). Second, corporate separateness from government is formally established (board no longer dominated by government representatives, debts to government annulled, any work for government to be contracted for). Third, the company may be transferred to private owners. The stages are not necessarily distinct.

The reform process in public marketing agencies in the case study countries is first overviewed and then discussed with regard to whether agencies can be reformed successfully and play a productive role in market development, or whether dissolution is the better option.

Zimbabwe[2]

Single channel marketing was the rule for the main agricultural commodities in Zimbabwe: grain, beef, cotton and dairy products. All had developed during the pre- and post-World War Two era when the control of agricultural marketing by the state had become the norm, particularly in southern and east Africa. But by the 1980s these arrangements were increasingly seen as inefficient, with some agencies deeply in debt. Reasons vary, but there appear to be a number of common causes:

- government pricing policies which did not allow them to cover their costs. This was a particular problem for the grain marketing agency GMB
- the 'soft budget' statutory relationship between the agency and government under which the central government budget is required to finance their deficits whatever the cause
- managers and boards of directors with little autonomy or accountability, with parent ministries interfering routinely in decisions regarding purchases, investment, hiring and firing of labour, pay levels and borrowing
- inadequate internal financial control and external monitoring systems
- managers not rewarded according to performance
- boards of directors often without the expertise needed to run the enterprise.

The ease of reorganisation of marketing agencies under the Economic Structural Adjustment Programme programme (ESAP) in the 1990s varied according to:

- whether their product was regarded as an essential staple (maize particularly) or essential input for manufacturing (cotton). In both cases the previous pricing policy of the boards had kept their product prices below import parity, indirectly subsidising their consumers
- the capacity of the agencies themselves to be agents of change
- their debt situation

- whether market liberalisation reinforced their future profitability (cotton, dairy) or undermined it (beef).

Dairibord Zimbabwe Ltd (DZL), the former Dairy Marketing Board, provides the leading Zimbabwe example of successful privatisation of agricultural parastatals. It moved quickly to profitability after taking over the former Dairy Marketing Board in 1994, by cost cutting through restructuring, including putting in new management, closing unprofitable depots and putting milk roundsmen on to commission instead of salaries. A debt/equity swap converted long-term loans into share capital. It became a purely commercial company without any statutory responsibilities. At the same time the dairy processing sector was opened to competition. DZL remained 100 per cent government-owned until 1997 when it was floated on the Zimbabwe stock exchange, with 5 per cent of shares being offered to small holders, 10 per cent to commercial farmers, 15 per cent to the national investment trust, and 10 per cent to staff. The flotation was oversubscribed owing to the strength of Dairibord's prospects. Dairibord profits were up in 1997. Dairibord had about 85 per cent of the Zimbabwe milk processing market in the late 1990s. DZL is one of Zimbabwe's largest manufacturing and marketing concerns with seven factories in Harare, Chitungwiza, Bulawayo, Gweru, Kadoma, Mutare and Chipinge. The Chitungwiza milk powder plant has capacity to process about 300 000 litres per day and thereby absorb milk surpluses. Milk supplies come mainly from specialised dairy herds in the commercial farm sector. DZL produces a full range of milk products, from fresh milk through to yoghurt and ice-cream. Most sales are domestic but there are exports including to Mozambique, Namibia, Malawi and the Congo. Trade restrictions plus the developed local dairy sector have prevented exports to South Africa.

The restructuring and privatisation of DZL have created a profitable and debt-free core to the dairy processing sector in Zimbabwe. Further development of the sector should see stronger competition from existing or new processors seeking a larger share of the profitable market. DZL purchase prices for raw milk compared to sales prices (Z$4/Z$9.90 per litre in the late 1990s) are criticised as low by competitors; but they would have to convince farmers to shift their supply contracts from DZL, which will require farmers having confidence in the longer-term prospects of the competitors. Processing capacity is critical to competition and profitability in dairying, an industry which has suffered from overcapacity in many countries. The best outlook for increasing competition and versatility in the dairy industry for the benefit of consumers and farmers small and large, may lie in lowering trade barriers and transport costs for trade with Zimbabwe's neighbours, including South Africa. Zimbabwe's technical efficiency in dairying is high (Attwood n.d.: 3). It should therefore be well able to compete.

The organisational strength of Zimbabwe's dairy industry is reflected in the ease with which the key milk quality assurance organisation in the Zimbabwe government (Dairy Services) was commercialised and made ready for running on a joint government–dairy industry association basis, with savings to government (staff costs) and no loss of effectiveness. The details are in Chapter 11, where the case is presented as an example of successful commercialisation of a quality control service. Only departmental politics in the Ministry of Lands and Agriculture in the late 1990s prevented the shift of dairy services out of the Ministry from being completed.

The Cotton Company of Zimbabwe was set up in 1994 as a public but unquoted company with 100 per cent government ownership, following the decision by government in 1993 in line with its Economic Structural Adjustment Programme (ESAP) to liberalise the cotton subsector, thus ending the monopoly of the former Cotton Marketing Board (CMB). The CMB had been put in place in 1969 as a parastatal with a monopoly mandate, responsible for:

- purchasing all seed cotton in Zimbabwe, except that grown around Beit Bridge which was exported to South Africa because of transport convenience
- controlling the varieties of cotton grown
- delinting and selling planting seed
- ginning of seed cotton and marketing of the products (cottonseed for oil crushing, lint for spinning, linters for paper manufacture, planting seed sold to growers).

CMB developed the infrastructure to carry out these responsibilities, involving backward and forward linkages from ginning to farmers, manufactures and export markets. Prior to independence in 1980 the bulk of lint produced was exported and the bulk of cotton production was on commercial farms. It offered cotton growers a fixed price, supplemented by a bonus if earnings subsequently allowed it.

Two important changes in the post-Independence era were:

(a) The vigorous effort by government to give communal farmers access to agricultural markets, by providing credit, inputs and purchasing.

The marketing boards were the vehicles for this process, which brought marked increases in output from the more fertile parts of the communal areas. Communal areas now account for almost three-quarters of total cotton production. This has necessitated a process of relocation of ginning capacity.

(b) Protection was granted to domestic textile production in the form of controlled prices for cotton lint below the export parity for lint, thereby obliging cotton farmers to subsidise textile production. This stemmed from depreciation of the Zimbabwe dollar in the early 1980s which increased imported input prices to manufacturers, leading the industry to plead for protection. The bulk of cotton lint output was switched from export to domestic sales to the textile industy. This reduced profits in cotton production and is said to have discouraged innovation in textile firms necessary to keep them internationally competitive – with severe consequences for them after the cotton industry was deregulated in 1993 under the Government's Economic Structural Adjustment Programme (ESAP).[3]

Cotton production peaked in 1989 at approximately 280 000 tonnes, reducing thereafter as a result of disincentives to farmers from the unattractive prices and drought in the early 1990s. A diminishing volume of exports was bearing the weight of the subsidy to the local textile industry. Cotton farmers lobbied to be allowed to export more and to escape from the restrictions placed on the Cotton Marketing Board. After liberalisation in 1993, they set up CotPro as an independent ginnery, with support from the commercial cotton growers association. CotPro also utilised the independent Triangle ginnery. During its reorganisation the Cotton Company of Zimbabwe (COTCO) sold off two of its ginneries to Cargill. The share of exports in total sales reverted to its pre-regulation dominance over domestic sales, i.e. about 70 per cent against 30 per cent domestic sales.

The commercialisation of CMB described above, from which COTCO was created with 100 per cent government ownership in 1994, was followed by privatisation through sale of 75 per cent of COTCO's shares in 1997. Restrictions were placed on the disposal of the shares as follows: 20 per cent were to be sold to small cotton farmers, 10 per cent to large cotton farmers, 5 per cent to employees, and 10 per cent to the National Investment Trust. Government retained 25 per cent, and the remaining 30 per cent were available to institutional investors and the general public.

Unlike Dairibord, whose share flotation was oversubscribed, much of the COTCO share allocation to farmers, employees and the National Investment Trust has not been paid up and the shares are being 'warehoused' by financial institutions. This has resulted in pressure from the financial institutions for a greater say on COTCO's board. Nonetheless, despite this setback privatisation has strengthened COTCO: it has helped to fund rationalisation and relocation of ginneries, involving retrenchment and new investment. Profits have been sufficient to allow a 5 per cent dividend in its first year of private status. It commands 70–80 per cent of the market for seed cotton from growers. As the privatisation prospectus notes:

> In the short to medium term, the company has a competitive advantage by virtue of its long established infrastructure, as well as both its long established and revolving inputs scheme and seed cotton contracts with farmers, which assure use of at least 50 % of the company's production capacity in any one year. The company has a wider network and substantially greater processing capacity than any of its competitors in the local market, and it will be difficult for competitors to increase their market share without further significant investment in their capital infrastructure.

Zimbabwe is a small player on the international cotton market, accounting for about 250 000 tonnes out of world production of about 19 million tonnes, in which the United States and China are dominant producers. Zimbabwe's cotton industry has used its small player status well, benefiting from an early orientation to exports which established high standards of quality control. Further, Zimbabwe's cotton lint is of particularly high quality, as a result of the good pre-ginning handling system reducing contamination and increasing uniformity, and the continued hand-picking of most of the crop.

As with dairying, investment in infrastructure is the critical entry barrier for new firms into the industry. COTCO's dominant position is based on its infrastructure (see above). Nevertheless, in the deregulated domestic and export market, expansion of the two minor players (Cargill 15 per cent, and CotPro 5 per cent) is more easily possible. The dependence on export markets also reduces the adverse effects that a degree of monopoly can have on price and quality of service.

In general, Zimbabwe's cotton industry has benefited in recent years from the much increased participation of communal farmers, and from deregulation and privatisation, despite the remaining challenge in the privatisation process to secure the transfer of shares to farmers and the National Investment Trust. Cotton production has been increasing once more since deregulation, though it has yet to regain its 1989 peak. The major shake-out in Zimbabwe's textile industry, which followed liberalisation of the domestic and external trade in cotton and in textiles, appears now to be increasing Zimbabwe's competitiveness in textiles (see Jackson 2002).

Less successful has been the corporatisation of the successor to the former Cold Storage Commission (CSC) after liberalisation of the domestic meat market in 1993. The Cold Storage Commission had been created in 1934 to market meat domestically and externally. The commissioners were mainly commercial farmers. It had a monopoly of meat supplies to the urban areas and of exports. In the post-war years it established an export market in UK, with the assistance of Commonwealth Preferences. Widespread cattle diseases during the civil war of the 1970s reduced exports. But after the end of the civil war in 1980 efforts were made to recover its export position: a levy-exempt EC quota was obtained under the Lomé Convention and investment in increased export-class slaughter capacity was undertaken, involving CSC in substantial borrowing on the domestic capital market.

However, this expansion came at the wrong time for CSC. First, CSC was about to lose market share severely in the domestic market, through market liberalisation.[4] Second, CSC ran into management difficulties in the early 1990s, with allegations of corrupt practices. Third, CSC (like GMB) has accumulated debts owed by government. Fourth, the interministerial committee on privatisation had yet to provide clear direction for the future of CSC, hampered (some argue) by lack of experience and high turnover of officials, interests of ministers, and the desire of government to determine to whom CSC would be sold or would collaborate with. CSC management wanted a business partner who is powerful and knowledgeable in the meat trade.

Together, these factors had left CSC with surplus capacity and virtually bankrupt by mid-1998. There had been some selling off of capacity in the mid-1990s and CSC wished to undertake more, but feared that would strengthen its competitors. It has recently sold off holding grounds on a lease basis. Having lost its legal monopolies, CSC's remaining markets have been exports to EU (its more up-to-date and hygienic slaughter facilities have EU veterinary approval, which its competitors presently do not have) and higher value local sales. Despite major refinancing by government the prospects for privatising CSC as a going concern were not good in the late 1990s. In sum, commercialisation of

CSC was poor, largely owing to unwise expansion and major mismanagement pre-reform. But the domestic meat sector is buoyant, making the fate of the former public agency of less importance, and strengthening the case for privatising it fully to cut losses.

The most challenging corporatisation task in Zimbabwe has been that of the Grain Marketing Board (GMB), the key player in the formerly closed, single channel maize marketing system. The Grain Marketing Board was the largest of the marketing boards reformed in the early 1990s, and the most critical for food policy. As the sole buyer and seller of maize, the key grain staple, its prices had been an important determinant of maize production and of the cost of living, and its stocks provided the main source of grain for current milling and against year-on-year shortfalls in marketed production relative to consumption. The maize economy of which GMB was the centre was tightly controlled: trading and small-scale milling (mainly hammer milling) of maize was outlawed in urban areas, thereby affording protection to large millers. The move to commercialise GMB fully in the mid-1990s was the result of accumulating GMB deficits. The deficits had been caused by:

- the successful post-independence effort to encourage communal farmers to market more maize, which involved GMB in extending its depot network, thereby incurring increased cost
- government's desire to keep down urban food prices and to pay pan-territorial producer floor prices, sometimes above market price level
- the distribution of maize for relief in the major drought of the early 1990s
- liberalisation of the domestic maize trade which produced competition for GMB in its most profitable activity (buying in accessible rural areas and selling in cities) but not in its most costly operations (buying and selling in remote rural areas).

GMB was given a new board with substantial non-government representation, management independence and commercial objectives, subject only to its carrying out non-commercial ('social') tasks for government, the costs of which would be fully reimbursed by government. These tasks included buying and selling from time to time at above or below market price (in order to provide subsidies to producers or consumers) and management of the strategic grain reserve. One immediate effect of the change was that GMB reduced its network of outlying, unprofitable depots, and reduced its employees. GMB market share fell quickly as the deregulated domestic buyers and sellers traded directly with each other.

By the late 1990s corporatisation of GMB had not yet produced the desired commercially independent, government-owned company. An immediate cause was the public expenditure crisis: with government deficits and debt at unsustainable levels, the resources and will to reimburse GMB immediately were not there. Further, government was reluctant to give up remaining levers of direct control over prices. Government continued from time to time to impose floor and ceiling prices, especially when prices were rising (e.g. late 1997), even though such price control was increasingly ineffective owing to GMB's declining share of purchases and sales (10–20 per cent of normal sales).

While government had agreed in principle to pay GMB for non-commercial tasks, in practice payments from the Ministry of Finance were often delayed (e.g. in 1997).[5] Temporary financial relief had been given in 1994 by government buying the strategic grain reserve from GMB, thereby cancelling GMB's debts, using the proceeds of a 5 per cent levy on taxpayers. For a year or two GMB managed to trade with a small profit. But imposition of floor and ceiling prices had helped to land GMB in a Z$183 million trading deficit in 1997, which government had yet to settle in mid-1998.

Interviews with GMB's management indicated that GMB's incentives could be distorted by its uncertain relation with government. For example, as the sole importer and exporter of maize, GMB was able to increase its profits from letting domestic prices drift up before releasing stocks. Further, if GMB were instructed to sell below market price it had little incentive to incur transport costs by moving stocks from surplus-holding depots to deficit areas, thereby reducing its effectiveness as a manager of the strategic grain reserve.

A riot in Harare in January 1998, motivated, according to the organisers of the protest, by high prices of food – but by political ambitions according to government – highlighted the tensions in Zimbabwe's food policy. Maize prices had risen sharply in the latter part of 1997, from $Z1400–1500 in October 1997 to over $Z2000 by the year end. Millers pressed the government for the strategic grain reserve to be released to dampen the rise in prices. When government did instruct GMB to sell the strategic grain reserve early in 1998 it obliged GMB to sell below market prices. But price impact was limited and government imposed price controls on grain products. This drew protests from the milling industry that the resulting margin between their costs and revenues made flour production unsustainable. Government in turn accused the milling industry of price fixing.

The tensions in Zimbabwe's food policy reflected the partial state of liberalisation of grain markets and of commercialisation of GMB:

- Maize prices in Zimbabwe rose above import parity in early 1998, suggesting that prompter imports would have stabilised the price. But GMB was the sole legal importer and could benefit commercially from higher local maize prices.
- Restrictions on maize flour imports fuelled the debate over pricing in the milling industry. Active cross-border trade in grain products would keep urban prices broadly in line with import parities.[6]
- The physical strategic grain reserve had become an inadequate substitute for external trade, because urban grain markets are increasingly large in relation to the physical strategic grain reserve. The strategic grain reserve's primary purpose should be for drought relief in remoter rural areas.
- If the commercialised GMB is instructed to sell the strategic grain reserve below market prices, yet has not been compensated by government for previous non-commercial tasks on behalf of government, it may have little incentive to sell off the strategic grain reserve vigorously.[7]

It was clear by the late 1990s that GMB's twilight status between state and private sector was in no one's interests. Moves were afoot to split GMB by setting set up a separate storage company (with a number of grain depots) to manage the strategic grain reserve. The rest of GMB's assets were to be fully privatised. This would have the advantage of freeing GMB's underused storage capacity for lease or sale to the private sector. It would also open up alternative options for managing the strategic grain reserve.

However, in 2000 political events overtook any further restructuring of GMB,[8] as President Mugabe embarked on land confiscation from the predominantly white commercial farmers, commercial maize production plunged and food prices escalated. In July 2001 government reimposed GMB's former monopoly on all trade in maize, and resorted to confiscation at sub-market prices of private grain stocks from leased GMB silos. Zimbabwe's burgeoning commodity exchange ZIMACE closed.[9] Rather than encouraging grain production, reimposing GMB's monopoly seemed likely to reduce plantings further, and was widely regarded as a means for government to secure maize supplies for its supporters before the 2002 presidential elections.

Sri Lanka[10]

The Paddy Marketing Board (PMB) was a statutory body under the Ministry of Agriculture, and historically one of the main government agencies in Sri Lanka's food security system. PMB was created in 1971 in order to undertake procurement for the country's rice ration scheme. In effect, PMB's initial role was as monopoly purchaser of the marketed surplus of rice, and between 1971 and 1977 private trade in paddy and rice was heavily circumscribed and discouraged. In order to undertake this role, PMB was endowed with substantial assets for the storage and milling of rice, comprising warehouses with a capacity to store 300 000 metric tons of grain and thirty rice mills. At the peak of its operations in the 1970s, PMB purchased around 40 per cent of the entire domestic rice crop.

PMB's functions and fortunes then declined, but in an uneven and unplanned way. The government which came to power in 1977 set about disbanding the rice ration scheme which had become budgetarily unsustainable due to its cumulative subsidy costs. The need for paddy purchases to supply this scheme declined and so, therefore, did the scale of PMB operations. The proportion of the domestic harvest procured by PMB moved erratically during the 1980s on a declining trend, culminating in a low point between 1989 and 1992 when less than 2 per cent of the annual rice harvest was purchased in each year. From 1993, this decline was reversed, and in the 1995 season PMB purchased some 10 per cent of total paddy production.

As part of its purchasing function, PMB implemented a minimum guaranteed producer price for paddy called the Guaranteed Price Scheme (GPS). In the 1970s, the GPS was set so that PMB was a relatively attractive outlet for paddy sales by farmers. In the 1980s and early 1990s, this situation was reversed and the market producer price of paddy was generally above the level of the GPS, even at peak harvest time. In 1993–5, there was another reversal, and the GPS was once again set at a level that made PMB purchases competitive by reference to the level of producer prices at peak harvest offered by traders to farmers. This was followed by the final policy reversal for PMB in 1996 when government denied it further funds to purchase rice, leaving PMB in a severely indebted state. It has been inactive since then though yet to be formally wound up. The fluctuating fortunes of PMB during its twenty-five-year history reflect vacillations in food security policy by government, in response to pressures from debt and donors to reduce expenditure, and pressures from rice farmers to raise prices by state purchasing. The liberalisation policy, first embarked on in the late 1970s, has been increasingly favoured since the mid1990s, despite evidence of hardship among rice farmers.[11]

As one of the main organisational pillars of Sri Lanka's food security policy, PMB had inter-agency links of varying importance with the Food Commissioner's Department, the Cooperative Wholesale Establishment, and the Multipurpose Cooperative Societies. Latterly, the most important link was with the Food Commissioner's Department, particularly via PMB membership of the influential Food Security Committee.

PMB also linked with its parent ministry, the Ministry of Lands and Agriculture, within which there is a specific strategic concern with self-sufficiency targets, producer price levels for rice, and their relationship to trends in the cost of rice production. The Ministry of Agriculture is also represented in the Food Security Committee, and it is therefore this forum which brings together, on the one hand, the domestic paddy production dimension of food security, and, on the other hand, the dimensions concerned with rice and wheat imports, wheat flour distribution, and wheat flour price levels.

As set out in the Paddy Marketing Board Act of 1971, PMB's legislative objectives were vague and all-encompassing (PMB Annual Report 1992). They comprise essentially two duties: (a) to carry on the business of purchasing, selling, supplying and distribution of paddy and rice; and (b) to carry on the business of milling, hulling and purchasing of paddy and rice. These duties appear to reflect the original intention to utilise PMB as a monopoly buyer of domestic paddy for the rice ration scheme, although this was not stated explicitly. From the late 1970s, PMB's role tended to be the more conventional agricultural policy objective of farm-gate price stabilisation.

It is clear that PMB was an unusual hybrid compared to other food grain parastatals in developing countries, and its lack of explicit policy objectives in the price and food policy arena rendered it continuously prone to political expediency under successive Sri Lankan governments. Typically an organisation of the PMB type might be expected to implement a floor and ceiling price policy, aided by domestic buffer stocks and control over the level of imports. This is the basic format followed, for example, by the Indonesian food security parastatal, BULOG, and, less successfully perhaps, by many food grain parastatals in Africa.

The ambiguous nature of PMB's objectives resulted from its origins in procurement for ration delivery rather than for price policy purposes.

Food security stocks have been handled by other agencies in the food system, principally the Food Commissioner's Department and the Cooperative Wholesale Establishment, with private bondsmen latterly playing a leading role (see Chapter 10 below). The PMB did not implement a price policy based on annual announced floor prices to farmers and target ceiling prices at the retail level. The Guaranteed Price Scheme was not announced annually, but intermittently, and the timing of changes in its level bears no relation to the planting decisions of farmers. From 1980–96, the Guaranteed Price Scheme was changed ten times, and the timing of the announcement of new levels occurred in seven different calendar months varying from January to November.

Assessing how well PMB performed is difficult, owing to the ambiguity of its objectives and the varying obligations which government placed upon it.

Effectiveness

Starting with a volume measure of its internal operations, the quantity of paddy purchased by PMB varied widely between 1980 and 1995, ranging from a high of 320 000 tons in 1983 to a low of 5000 tons in 1989. On its own this volume measure tells us little about performance because these fluctuations may have been imposed on the organisation by external events, both natural (the size of rice harvests) and governmental (decisions about Guaranteed Price Scheme level and finance available for crop purchase). Nevertheless, the extreme nature of this variability in level of operation by itself implies severe difficulties for internal operational efficiency due to the variability it causes in capacity utilisation of both human and physical assets.

Fragmentary evidence suggests that even in its limited role of guaranteeing prices, under variable degrees of government encouragement, the PMB was ineffective. Monthly price data shows that PMB failed to provide a floor to the market in the lowest price months. Farmers reported that they would never deliver to PMB directly due to failure to pay in cash, uncertainty about whether money owed would be paid, and inconsistency in the application of quality criteria by PMB purchasing officers. Therefore, private collectors intervened between farmers and the PMB, and farmers did not get the floor price at farm-gate level.

The sources of PMB's weaknesses were not addressed in over two decades after economic liberalisation was initiated in 1977. It suffered from unpredictable reversals in government policy, and lack of explicit and consistent objectives.

Efficiency

From the mid-1980s, PMB underwent a considerable contraction in the size of its permanent assets, both physical and human. By the mid-1990s it owned four mills, as compared to thirty previously, and 166 storage warehouses capable of storing about 180 000 tons of paddy, compared to 300 stores previously. In 1989–90 permanent PMB staff levels were reduced from 1500 to 870 persons, subsequently falling to 500 persons in 1995. This downsizing was financed by central government, under an early retirement programme implemented across the civil service. In an incidental sense, the contraction of PMB staff implied a large increase in physical productivity per person. The quantity of paddy purchased per permanent employee rose from a lowest level of 3 tons in 1989 to 569 tons in 1995. It would be a mistake, however, to infer a rise in organisational efficiency from this trend. The picture is incomplete (it excludes use made of temporary personnel) and the rise in productivity is entirely driven by external exigencies (the government retrenchment programme and the reversal of policy on paddy purchases). The increased paddy purchase programme from the late 1980s proved financially unsustainable and in no way resulted from better resource use by PMB. An examination of the financial position of PMB provides further evidence of its inefficiency. Data for PMB's profit and loss accounts in 1992 and 1993 suggest that it made modest gross trading profits on its rice trading and milling activities but losses on rice noodles production and other operations. Owing to its substantial overhead costs, its net losses amounted to 15.5 million rupees in 1992 and 16.46 million rupees in 1993. Had provision for bad debts been made these losses would have been even greater, amounting to 21 million rupees in 1993 (Harrison 1995: 39). Thus, at the end of 1993 the PMB was already in a very weak financial position. A significant deterioration occurred in 1994 and 1995 as in both these years the PMB was active in the market buying paddy at prices above those paid by private traders.

Sustainability

The scale of PMB operations was related more closely to the finance for crop purchase allowed by government than to the technical implementation of the Guaranteed Price Scheme as a floor price for paddy. The sudden reversal in the volume of PMB purchases from 6000 tons in 1992 to 280 000 tons in 1995 corresponded to a change in the government in power and an instruction to the state-owned People's Bank to make sufficient finance available for PMB to exercise fully the GPS of Rs 7.42

announced in May 1993. The rise in paddy purchase operations resulted in a cumulative debt of the PMB to the People's Bank of Rs 2.5 billion during 1993–5. By 1996, the PMB had no means of paying the interest on its loans, let alone repaying the accumulated principal. Its operation under the market circumstances of 1993–5 was loss-making by definition, since purchases at the GPS were matched by sales in a rice market depressed by record harvest levels and the simultaneous advent of a large subsidy on the retail price of wheat flour to consumers. The ability of PMB to continue incurring such heavy losses was highly constrained. Although on paper its total assets for 1993 were valued at 396.9 million rupees, considerable doubt existed regarding its ability to collect old debts and realise its assets. It was estimated in 1995 that PMB's equity amounted to no more than 10 per cent of total assets (ibid.: 40). Government's subsequent closure of PMB seemed motivated as much by avoidance of further debt as by any change of direction on food security policy.

In summary, the targeted food stamp programme with which government replaced its universal rice ration in 1979 did not require state purchasing and sale of rice. PMB then appears to have been left to wither on the vine, as has been done in some other adjusting countries (e.g. Uganda after trade liberalisation in the early 1990s) with public marketing agencies which no longer served a purpose but were politically embarrassing to close immediately. With Sri Lanka's rice farmers protesting against falling real prices of rice in the early 1990s votes were won by promising to support farm-gate prices. PMB was used for the purpose by the new government, but this was not sustained after 1996.

Ghana[12]

Public marketing agency reform in Ghana appears to go against the pattern elsewhere: reform of food grain agencies has been less complex than that of non-food grain agencies. Reasons for the difference include:

- the diversity of food sources (root crops as well as grains) on which Ghanaians depend, compared to the one or two staple grains in other case study countries (maize predominantly in southern and eastern Africa, rice and wheat in Sri Lanka and India)
- the long-term dominance of private firms in Ghana's food-grain trade

- the critical role of cocoa and the state-owned conglomerate Cocobod in Ghana's economy, making reform complex. 'Cocoa is still vital to this country from a macro perspective. It contributes 13 to 14 per cent of GDP, 11 per cent of tax and 30 to 35 per cent of foreign exchange earnings. As an economic activity it embraces about half the population and is grown in six of the 10 regions. It is important for government to pace itself [in privatising Cocobod]' (John Newman, chairman of Cocobod).[13]

Food grain agencies

Reform of food grain agencies was not included in donor conditionality in Ghana's structural adjustment programme and only came on to the reform agenda in 1990. Reform involved closing the Grain Warehousing Company and abandoning the guaranteed minimum price (GMP) for maize which the state-owned Ghana Food Distribution Corporation (GFDC) had nominally run to stabilise farm-gate prices. Ending the GMP saved public funds which had subsidised GFDC, but made no impact on food grain markets since by 1990 the GFDC handled only about 1 per cent of maize traded and no rice.

The GFDC has the most advanced drying and storage facilities in the grains market in Ghana. Its storage capacity was estimated at about 75 000 tonnes (about 10 per cent of average maize output between 1988 and 1992). Its staff were also considered to be technically capable in the handling of maize stocks (Danida 1993). But by the mid-1990s utilisation of GFDC's drying and storage capacity had fallen to about 5 per cent (Danida 1993). Failure to wind up or sell GFDC was partly to blame, since GFDC was paralysed by debt and unable to trade. Efforts to improve credit delivery to maize traders via a warehouse receipts scheme (see discussion in Chapter 8) were hindered by the incomplete reform of the GFDC, since ownership of grain stored in its silos could not be guaranteed (Coulter and Shepherd 1995).

The dismal performance of GFDC was the result of poorly defined corporate objectives which combined private commercial grain trading with public food security responsibilities and supplying soft-budget public sector organisations like schools, military, prisons and hospitals, which constituted the bulk of the corporation's customer base. Failure to pay by public sector organisations, and by government for disaster relief supplies, resulted in accumulation of huge bad debts. This, in turn, had weakened the corporation's capacity to meet its debt obligations, leading to its poor credit rating in the banking community, and limited access to credit for working capital. Restructuring of the GFDC involved divesting its cold storage and rice milling units (World Bank 1992). Further restructuring required laying off 25 per cent of its staff, cutting back further on its non-core activities and creating separate accounting for the management of government food security operations. But implementation of this plan was blocked by lack of funds, particularly to meet redundancy payments. GFDC has been non-operational since the mid-1990s.

Cocoa marketing agencies

The cocoa export industry has dominated the reform agenda of government and donors since structural adjustment began in 1983. Reform debates have focused on liberalisation of internal and export trade in cocoa, with transformation of the agencies grouped under the state-owned monopoly Cocobod a secondary consideration, but nevertheless tied up with the process.

As outlined in Chapter 5, reforms since 1983 include drastic cutting of Cocobod staff, licensing of private buying companies, floating of a majority share in PBC to producers and private investors in 1999, as well as opening of 30 per cent of the export market to licensed exporters for the 2000–1 season. Other undertakings given by government to IMF in 1999 were to merge Cocobod's extension service into the government's service, and to give equal access to credit and warehousing to all the licensed buying companies and PBC.

Here we review the debate and process of reform in internal and external trade in cocoa and the implications for the future role of Cocobod.

Reform of external trade

The World Bank has been a driving force behind the reforms associated with structural adjustment in Ghana. It is now pushing for full liberalisation of the cocoa market, particularly the external trade for which Cocobod's Cocoa Marketing Company (CMC) held a monopoly until 1999 when 30 per cent of sales were allocated to licensed private companies. The major argument for full liberalisation is that this will enable the improvement of returns to producers by reducing marketing costs. With government now committed to expansion of cocoa production by raising producer prices, liberalisers argue that government is paying more to support Cocobod than is being gained from the premium on Ghana's cocoa; and that government should instead use the resources to improve infrastructure for farmers, upgrading feeder roads to marginal areas, reduce taxes and encourage value-added processing.[14]

However, experience in Nigeria, Cameroon and Côte d'Ivoire indicates that export liberalisation increases the producers' share of a reduced f.o.b. price. The country may lose in a number of respects: prices for cocoa deteriorate, so less foreign exchange is realised; contracts are no longer guaranteed in the same way; overall quality control disappears, reducing the premia paid; and the futures market disintegrates. But it gains if overall production increases in the medium to long term, as is now Ghana's strategy. If export liberalisation benefits cocoa farmers, higher farm incomes would be expected to lead to greater investment in cocoa farms, and a higher rate of growth of agricultural and national product, with important spin-offs for local economic development in the cocoa-producing areas, more prosperity along the trade routes, and higher total revenues for government, assuming liberalisation did not damage its ability to collect the tax.

The key risks are: can Ghana liberalise its exports without losing quality and markets? Will export liberalisation and higher prices for farmers deliver an increase in supply? If so, at what cost to forests, since cocoa expansion in the past has involved destruction of forests? And will the tendency of liberalisation to raise production of a commodity with limited elasticity of demand[15] reduce overall revenue to the industry?

The strongest argument against rapid and full liberalisation of Ghana's cocoa exports has been the strong performance of CMC in its export markets. Consulted in the mid-1990s, the vast majority of customers of the CMC were extremely happy with its performance, according to a study of further liberalisation potential carried out for World Bank (LMC International 1996a). Over forty cocoa buyers were covered by the survey; all declared their satisfaction. Further, only a few commodity traders were looking to enter the Ghanaian export market. The CMC was seen as being 'in a different league' to the Caisse in Côte d'Ivoire (now defunct) or that in Cameroon. It had a 'triple A' credit rating among international financing organisations – having never defaulted on loans. It fulfilled contracts and its dealings were straightforward – with no evidence of corruption. Its staff were well trained and competent. It did not suffer political interference. Its marketing costs were small, and its use of the countervailing power available to it as the controller of the major bulk premium quality cocoa on world markets enhanced and stabilised the prices it has received.

CMC sells in the London futures market, and unlike other cocoa producers is able to sell cocoa up to a year in advance. This impressive performance from a public sector commercial undertaking is combined with the fact that Ghanaian cocoa is of the highest quality, as a result of quality standards maintained by the Quality Control Division (QCD) of Cocobod. QCD's service is costly (revenue from rejected beans is lost) but effective. The cocoa coming from countries which had fully liberalised their exports, by contrast, sells only through spot markets, and at a discount to Ghanaian cocoa.

Promoting competition in internal trade

In 1992, as a result of continuing dissatisfaction over the low proportion of f.o.b. prices received by farmers and high costs of marketing,[16] licensed buying companies (LBCs) were allowed by Cocobod to compete with Cocbod's Produce Buying Company (PBC) in the domestic purchase of cocoa. By the 1995/6 season eight LBCs were estimated to be buying 35 per cent of the crop.

The process of introducing competition into this market has been gradual. LBCs have been supported by the Cocobod with information, a substantial line of credit, and in some cases preferential access to the 'tools of trade' (e.g. sacks, weighing scales, warehousing). Licensing generally proceeded smoothly, without any obvious favouring of any one political group in allocating licences. But although many companies have been licensed, few have yet traded seriously. This may be because there are many practical difficulties for firms to surmount before they can seriously engage in cocoa trading. But there may have been strategic reasons for registering as an LBC, e.g. to be in the fray once liberalisation of external trade took place.

Performance of the Produce Buying Company, Cocobod's internal purchasing arm, has been much less impressive than that of its export arm, the Cocoa Marketing Company. Change has been slow, despite introduction of competition via the licensed buyers in 1992. In the mid-1990s, the Produce Buying Company (PBC) was not at all autonomous, and was not operating on anything approaching a cost-covering basis, although payment for some services provided by other parts of the Cocobod (e.g. the Quality Control Division) had been introduced. Despite its advantageous access to warehousing and finance, and therefore being able to undercut its competitors, PBC's performance has not been strong. It had to cut staff substantially, from 16 000 in 1990 to 2500 in 1995, with further plans for reduction. From 100 per cent of the market in 1992 it was down to 65 per cent by 1995/6. Only the financial weakness of the LBCs prevented further loss in the 1990s. The problems faced by the PBC were largely due to the lack of an adequate and clear framework within which to operate as a commercial entity. Despite its weaknesses, PBC remains the effective price-setter in the trade. Under IMF pressure government decided in 1999 to privatise PBC. Given the disappointing performance of PBC it is not surprising that the offer was undersubscribed.[17]

Future role of Cocobod

The options for further reform in Ghana's cocoa marketing were laid out by LMC International (1996a: Chapter 5) in a review sponsored by the World Bank. These were:

1   maintaining the status quo;
2   a reduced role for Cocobod but maintaining the Cocoa Marketing Company's export monopoly, with further privatisation of internal marketing, storage and quality control, a reduction in export taxation, and tying the minimum producer price to a share of the f.o.b. price;
3   a share of cocoa exports to be handled by the private sector, limited by quota or number of companies licenced to export;
4   complete liberalisation.

In its reform programme, government has opted for elements of two and three, to enable CMC to pay a predetermined producer price (67 per cent of f.o.b. price is aimed for), to reduce export taxation, and to give time for private buyers to strengthen their performance. As for Cocobod itself, the view in the industry in 1999 was: 'The Cocobod will be restructured to provide advisory and regulatory functions. It will formulate guidelines that the private sector must follow.' All its agencies which participate in markets will be sold off or merged with Ministry of Food and Agriculture services.[18] But such changes are unlikely to be rapid. To offer at least 67 per cent of f.o.b. price to farmers, government control of prices is needed. To ensure quality, the task of switching from QCD's active intervention to a regulatory role enforcing minimum standards on private operators has yet to be tackled.[19] Arguments that the quality control function is redundant since private buyers will purchase the quality they need (Gilbert, Varangis and ter Wengel 1999) conflict with government policy which is to defend the reputation for high quality which Ghanaian cocoa enjoys. Further, crop finance for small producers for small producers will need to be operated satisfactorily by private companies.

In sum, Cocobod is a rare example of a substantially successful public export marketing agency based on smallholder producers. It seems that even if CMC is privatised in part, government will be unlikely to let go quickly of its direct controls over the industry. This in turn may slow down the conversion of a restructured Cocobod into a regulatory agency for a privatised industry, or into a representative organisation for the industry.[20] The liberalisation debate regarding Ghana's cocoa exports is far from over. The underlying strategic questions remain: maintaining quality and finance to farmers in a liberalised trade; whether expansion of cocoa production from existing farms is feasible; how to ensure that expansion does not destroy forests; whether the impact of large increases in production on producer prices will be negative; and whether falling prices, together with low yields, will drive smallholders out of production – which would have heavy social costs.

Kenya[21]

Reform of the National Cereals and Produce Board (NCPB) has been at the centre of maize market liberalisation efforts in Kenya since the early 1980s. Despite formally liberalising internal trade and maize imports (1993), formally ending the state corporation status of NCPB (1996) and setting it on a commercial footing, NCPB is substantially unreformed. It remains the instrument of government for intervening in grain markets, exporting surpluses, and managing a strategic reserve.

Furthermore, the underlying problem with Kenya's maize market remains: white maize is both the principal staple food and the principal crop of Kenya's commercial and subsistence farmers, and it relies on erratic rainfall. Widely varying outputs and prices result. Bumper crops (as in 1994 and 2001[22]) drive down prices and farmers plead for action by government to buy up the crop. Droughts (as in 1997) drive up prices and produce hardship for semi-subsistence farmers and poor consumers, resulting in efforts to distribute subsidised maize to the needy. Hence NCPB's former monopoly role: to stabilise prices to farmers and consumers, using buffer stocks and some exports and imports as the means to do so. But the heavy fiscal costs of this could not be fiscally sustained, resulting in unmanageable debts to NCPB. Further, NCPB's monopoly position was widely perceived to encourage fraud and waste, driving up its costs.[23]

The Cereals Sector Reform Programme (CSRP) was designed mainly to increase opportunities for private trade, by liberalising domestic and external grain trade in a phased manner, and by reducing NCPB's purchasing and sales network. The secondary objective was to reduce accumulation of public debt in NCPB. NCPB's public role would be to manage a small strategic grain reserve on a contractual basis for government. Prices would increasingly be set by trade, and reflect external prices. In this way private trade was expected to bear the risks and reap the profits of grain movement and storage, which should help to stabilise prices.

Implementation of CSRP revealed its limitations: NCPB's accumulated debts were written off at the start of the CSRP in the late 1980s. But by 1993 its balance sheet had deteriorated once more, due to fewer subventions from government, non-payment of debts by government departments, and the Board's financial indiscipline (see Figure 7.1). Therefore the Board fell back into the familiar pattern of late payments to farmers. The farmers increasingly turned to direct sales to millers, and the board lost market share through the mid-1990s.

NCPB's capacity: management reorganisation at NCPB was intended to improve the calibre of senior management through external recruitment, staff training, higher salaries and retrenchment. Recruitment of senior staff was necessary following the mass sacking in 1988 for alleged corruption.

However, a new organisational structure was slow to be defined, the Board resisted recruiting externally, and staff numbers were higher after CSRP than at the beginning (CSRP Monitoring and Evaluation Reports). The Board did not provide counterparts to understudy expatriate staff in key areas such as planning, information, transport and financial management. Accounting and financial management improved during the time expatriate staff worked at NCPB. But when they left the systems could not be sustained because there were no skilled people to take over.

The capacity of NCPB remained severely limited in the mid-1990s, as the summary of constraints indicates in Figure 9.1.

Evidence suggests NCPB has long suffered from political interference, reliance on subventions from the Treasury in order to sustain prices at the levels policy demanded, failure of the Treasury and government departments to pay for services, and resultant inability to support market prices in line with price movements, or to sustain a floor price for maize farmers. The financing problems resulted in inability to prepare budgets linked to revenues and needs. There was unprofessional recruitment of staff, low morale of staff, excessive centralisation of decision-making in the office of the managing director, governance by a board of directors which was mostly politically appointed and did not have adequate representation of farmers and was slow to adjust to changing circumstances. These problems continued into the 1990s (Lewa 1995; Lewa and Hubbard 1995).

INTERNAL CONSTRAINTS

Financial resources

Inadequate financing. Inability to prepare budgets linked to revenues and needs. Vulnerability to budget cuts by Treasury. Inadequate subventions for social roles. Financial mismanagement. Large outstanding debts by government bodies. Inability to collect debts from debtors within government.

Human resources

Excess staff numbers in the lower ranks. Unprofessionalism in operations. Poor salary structure pegged on civil service pay scales. Excessive centralisation of decision-making in the office of the MD. Politically appointed board of directors leading to influence of decisions by politicians. Inadequate working equipment. Little exposure to computer systems in operation. Low morale of staff. Poor staff training. Senior staff appointments influenced by politicians. External recruitment resisted by top management. No clear sense of mission by the leadership. Senior staff removed from the rest. No free interaction among staff. Generally poor human resource management.

Organisational autonomy

NCPB not independent of government. Board cannot set its own decisions without government influence. Staffing levels influenood by government. Little contact with all maize industry players. The public views NCPB as extension of government.

Management and administration

Managers not very clear of their own roles. Insecurity and fear that one can be fired any time. Political appointment of MD, Board's Chairman and Board of Directors. Poor delegation. MD has to vet all decisions before implementation. Subordinates have little trust of their bosses.

Commercial orientation

Poor response to the needs of the market. Poor relationship with consumers of the Board's services. Consumer orientation lacking.

EXTERNAL CONSTRAINTS

Excessive political interference. Failure to plan future operations independent of government. Reliance on inadequate Treasury subventions. Government decision to maintain NCPB as its agent for food security roles. Control by powerful political interest groups. Central government influence on procurement of maize. Market liberalisation and growth of competition reduced NCPB earning opportunities.

Source: Lewa 1998.

Figure 9.1 Summary of key constraints on NCPB

Network rationalisation to be undertaken under CSRP required closing a percentage of local buying centres and selling or leasing depots. Although many LBCs were closed, depots increased in number. Selling or leasing depots was not politically feasible.

Transport rationalisation required developing the Board's forecasting capacity, and improvement to the railway system particularly through investment in rolling stock and locomotives, while funding the Board to acquire its own custom-built wagons. This never happened, as the EC withdrew its support for several components of CSRP.[24] The Board has not been able to use the supposedly cheaper, bulk transport by rail because of the run-down condition of the railways.

A crop purchase revolving fund (CPRF) initiated by the EC to improve the Board's cash flow initially helped in paying farmers promptly, funding operating losses and competing for maize supplies with the private sector. But the CPRF was introduced without any clear protocol. It comprised part of the Board's finances, was not accounted for separately, was not used only for crop purchases as envisaged, and there was no mechanism for enforcing its correct use. It was soon leaking like a sieve.

NCPB's role as buyer and seller of last resort was poorly defined in CSRP. The government agreed, after intense pressure from donors, that the NCPB be limited to managing a strategic reserve of maize of a maximum of three million bags and that it buys at no more than export parity and sells at no more than import parity (Government Policy framework paper for 1994–6). Government further agreed to create a $60 million foreign exchange fund, enough to buy another three million bags to cover any emergency imports (ibid.). However, since the liberalisation of the maize market in 1993 government has mandated the Board to buy maize above export parity, and has not restricted NCPB to the maintenance of strategic reserves of three million bags only. Government seems determined to maintain a role for NCPB in price stabilisation, in addition to maintaining the strategic reserves. This amounts to NCPB's previous, unaffordable role which the reforms were designed to change.

The main achievement of CSRP was liberalisation of the domestic maize trade and maize imports. This was a groundbreaking change, changing the opinion of many who had doubted that the private sector could finance and deliver imports (Lewa 1995). Import liberalisation was facilitated by convertibility of the Kenya shilling, introduced at the same time.[25] Except for liberalisation, CSRP relied on reduction of state involvement in maize marketing as the means to stimulate increased private sector participation. Some feel this was a design error.[26] However, Kenya's political economy in the last decade has not favoured market development, so CSRP components dedicated to assisting private sector development may have been no better implemented than others.[27]

Import liberalisation has made maize supplies more readily available to consumers, particularly in the coastal area around Mombasa, and along the borders with Uganda and Tanzania, both of which export grain to Kenya. But it does not appear to have stabilised prices, particularly for grain producers, who protest vigorously that imports frequently glut the market (e.g. in 1994, 2000 and 2001). The underlying problem is that Kenyan maize production is not competitive with imports.[28] Government's use of NCPB to raise prices to farmers, and unpredicatability of the variable import tariff, may also discourage traders and millers from contracting with farmers. Further, trade liberalisation has aided the rise of small hammer millers (*posho* millers) and reduced the market share of large scale millers (Lewa 1995). Together, these factors may have resulted in trade liberalisation distancing traders and millers from grain farmers, rather than bringing them into closer, forward purchase arrangements. Long-term solutions will involve raising farm productivity, and switching to higher value crops.

In sum, the maize trade has been liberalised while NCPB continues to be used by government to intervene substantially to increase prices for maize farmers. This does not encourage market development in maize. If government leaves NCPB with large debts, it is likely to cease trading, as happened in similar circumstances to Sri Lanka's Produce Marketing Board (PMB) and the Ghana Food Distribution Corporation (GFDC) in the mid-1990s.

India[29]

The broad thrust of Indian policy since 1990 has favoured liberalisation and privatisation. Trade in manufactures and cash crops has seen considerable market development as a result. But food grain marketing has been an exception to this trend, particularly domestic marketing. The pillars of India's food security policy remain in place: government investment in agricultural technology, infrastructure and services to ensure availability of food grains; and a public distribution system for grains to ensure access by all. The role of the Food Corporation of India (FCI) is the same as the PMB in Sri Lanka before reforms: to procure grain at prices determined by government for distribution as rations.

Calls for reform of FCI and food grain distribution in India are motivated by waste (in grain storage particularly), rising costs of FCI (management, subsidised credit), suppression and crowding out of the private sector (excessive regulation, limits on use of agricultural produce as collateral, prohibition of futures trading in grains, restriction of milling profits, underinvestment in market places), and ineffectiveness for benefiting the poor through the network of over 400 000 fair-price shops (unavailability, fraud, prices too high, too much going to the non-poor) (Umali-Deininger and Deininger 2001).

The main attempt since 1991 to reform India's Public Distribution System is the Targeted PDS introduced in 1997. This attempts to target more food grain to people below the poverty line, raising the ration to 20 kg per household month. The price of food grain distributed to households above the poverty line is raised to market level.

Further reforms proposed by the World Bank are threefold. First, better targeting and delivery – by pilot food stamps distribution by private firms, phasing out all rations for the non-poor, using open market sales to maintain price stability between and ceiling and a floor instead of fixed prices instead of fixed prices. Second, promote market development – FCI buying and selling at market prices, developing warehouse receipts as credit collateral, removing the ban on grain futures trading, use competition policy instead of trade suppression to prevent monopoly, investment in market infrastructure and promoting market information. Third, modernise FCI – by shifting towards a financing and coordinating role, subcontracting market operations to private companies, operating with a hard budget constraint (ibid.).

Is FCI in a better position to survive reform and to contribute to market development than the grain marketing parastatals in Sri Lanka, Zimbabwe, Kenya and Ghana? Switching entirely from rations to a food stamps programme (as in Sri Lanka) would remove FCI's grain procurement and distribution role, since stamp holders would buy directly from markets. It is doubtful whether FCI would be better suited than a welfare department to manage the finance and implementation of a targeted income transfer programme that would have little or nothing to do with grain management. FCI already contracts out some of its operations (e.g. storage, see Chapter 10). The major contribution to market development is likely to come from deregulating domestic trade. FCI could only survive imposition of a hard budget constraint if it is allowed sufficient margin between buying and selling prices to meet its costs. In a deregulated and increasingly competitive domestic market this margin would narrow, requiring reduction in costs. Managing the strategic grain reserve would provide an ongoing role for FCI. But market development would reduce the size of strategic reserve needed and open up alternative means for managing it (e.g. privately owned stocks in bonded warehouses as in Sri Lanka). In sum, while FCI might survive liberalisation it seems unlikely to contribute to market development.

If trade barriers are lowered and subsidies reduced, the future politics of grain markets in India may revolve around how competitive are India's grain farmers, as happened in Sri Lanka and Kenya. In Sri Lanka the Produce Marketing Board (PMB) was set up with the same primary purpose as India's FCI: to procure grains for distribution as rations. After universal rations were replaced with targeted food stamps in 1979 PMB's role declined, but it was revived in the early 1990s to purchase rice in order to raise farm-gate prices, the same role that NCPB in Kenya has taken on. In both countries the underlying problem was farmers' difficulties in competing with imports.

However, the determination with which India has built up its agriculture over the long term, its strong economic growth, and its willingness to pay FCI deficits, suggest that it will not be forced by financial crises and consequent donor pressure into liberalising so radically that its grain production cannot compete.

Agency type and competitiveness of product

Some common features emerge from the case studies, supported by literature on other agencies and countries. In general, reform experiences differ according to whether the agency is marketing grain or other produce, and whether the produce is competitive with imports.

Type of agency: grain or non-grain

In general, reform of non-grain marketing agencies has been more purposeful, rapid and successful than reform of public grain marketing agencies. Liberalisation of dairy sectors has stimulated growth in Zimbabwe and Kenya. The privatised dairy agency in Zimbabwe (DZL) has prospered and expanded the market. Similarly, privatisation of the cotton marketing agency in Zimbabwe, while encountering some difficulties in raising capital from smallholders, has helped develop the market. In Ghana, Cocobod (still in state ownership but with reduced staff and subsidiaries) is playing a role in developing competition in both domestic and external marketing of cocoa. Where reform of non-grain agencies has failed, mismanagement and political interference have been the reason, preventing their privatisation as going concerns (e.g. attempts to recover the viability of the state-owned Cold Storage Company of Zimbabwe).

By comparison, efforts to reform grain marketing agencies (while maintaining them as public bodies) have generally been unsuccessful, owing to uncertainty in grain marketing policy and lack of an essential role for a public grain marketing agency in a liberalised market. Where public grain marketing agencies were prominent (e.g. Kenya and Zimbabwe) early reform efforts paid much attention to strengthening their management and focusing their objectives, in addition to liberalising trade. These involved:

- Corporatising the agency, to give it greater distance from government and reduce political interference. This involved making it statutorily answerable only to the highest level of government or to Parliament, drawing the majority of its board members from outside government, with the objective of covering its costs.
- Writing off its debts, which were usually the result of non-payment by government bodies for grain received or distributed on behalf of government.
- Government to pay for all services demanded from the agency.
- Downsizing the agency and selling off or hiring out surplus assets (e.g. buildings, silos).
- Investing in upgrading the management and systems of the agency.

In retrospect, these agency reforms lacked direction. They aimed only to create a more efficient and accountable public agency without specifying clear objectives or limits. This lack of direction reflected policy uncertainty regarding liberalisation, particularly how much of a future role government would want to retain – in external trade, price stabilisation and grain distribution for relief. Experience has shown that where liberalisation policy is uncertain, the agency has been used to interfere inconsistently in markets, for short-term objectives (e.g. raising farm-gate prices), and has even been allowed to exploit any monopoly power it has to increase its own revenues (e.g. Zimbabwe's GMB allegedly delaying maize imports in order to raise domestic prices for its sales in 1996). Further, when government is under financial pressure, old habits of failing to pay agencies for services demanded reassert themselves (e.g. Kenya). As a result, the reformed agencies have remained weak and still subject to political whim. None of this favours market development. Where governments have adopted a clear policy direction on liberalisation (e.g. UK, Australia, New Zealand, South Africa) grain marketing agencies have been quickly privatised rather than any effort made to reform them.

Lack of an essential public role for public grain marketing agencies in liberalised markets contributed to the lack of direction in efforts to reform them. The future role typically assigned to grain marketing agencies undergoing reform is to manage a reduced strategic grain reserve, with the purpose of reducing extreme fluctuations in price, or for relief supplies. But this is a service which any grain handling company with access to storage can deliver on contract to government. Moreover, alternative ways of providing the strategic reserve are available, such as bonded private stocks (Sri Lanka) and earmarked financial reserves, which avoid public management of grain stocks.

Policy and planning, to reduce transport and financing bottlenecks and to increase competition (e.g. Sri Lanka's Food Commissioner's Department in Sri Lanka, and food policy committees in government more generally) is a public role which is expanded by liberalisation of grain markets. Targeted welfare relief may also be an expanding public need. But the skills of public grain marketing agencies are confined to grain handling, transport and storage. There is no clear promotional role, since grains are graded commodities, with most usually consumed domestically rather than exported. Any public role in quality control/food safety inspection for grains would use only a fraction of agencies' resources. Marketing agencies' weakness as business entities usually means they are not viable for privatisation as going concerns, causing them to be sold off for the value of their assets only (mainly buildings and silos) when government decides to privatise them.

Competitiveness of the product

In general, reform of institutional arrangements in marketing has been easier and more successful for commodities which a country produces competitively compared to imports. Where production is buoyant and competitive, governments are spared the political pressure from farmers to raise prices when trade liberalisation exposes them to competition from imports. Consequently governments' use of public marketing agencies is more consistent. But in Kenya and Sri Lanka grain production is higher cost relative to imports. Protests from grain farmers against imports and low prices resulted in the use of NCPB and PMB to buy up grain, thereby increasing their debt, contrary to the policy of reducing their role and debt. By contrast, liberalisation of rice marketing in Vietnam, a competitive producer of rice, was managed without disruption, stimulating production so that both exports and domestic consumption increased (Minot and Goletti 2001). In Bangladesh, a rising trend of rice production and private trade in rice assisted reforms including trade liberalisation, abolition of universal rations and restriction of the role of the public marketing to managing the strategic grain reserve (Ahmed *et al.* 2000). Similarly, in South Africa, liberalisation of domestic and external trade in maize in the mid-1990s was accompanied by rising production and constant consumer prices for maize meal (Bayley 2000). This raises the prospect for China and India – both with rising trends of grain production – to experience an easier transition to freer trade in grains than if it threatened domestic grain production. Indonesia, where rice prices have been supported above import levels,[30] may find the transition more difficult.

Reform or dissolution?

In the light of experience, do public marketing agencies have a continuing role to play in facilitating development of markets for agricultural products? Should commercialisation of public marketing agencies be part of a reform strategy or should they be dissolved? There are three main observations from this review.

First, public marketing agencies usually have little continuing public role to play in a liberalised market. Among our case studies only Cocobod in Ghana has a continuing vital role, owing to its quality control function, and its strong reputation in export markets. Only in such circumstances are efforts to reform an agency for a future public role likely to be fruitful. Plans are to transform Cocobod gradually into a regulatory and planning body for the industry.

Second, incomplete privatisation of public marketing agencies reflects uncertainty in policy and can be destructive to market development. This is more likely with grain marketing agencies, particularly where grain production is less competitive with imports. Sri Lanka made little effort to reform its grain marketing agency (PMB), but used it for short-term political ends to raise prices for rice farmers in the early 1990s. In Kenya and Zimbabwe commercialisation of the grain marketing agencies (NCPB and GMB) was prominent in early reform packages, but they too have subsequently been used for short-term political ends, at the expense of market development. Among developing countries, Bangladesh's experience – in which a clear policy direction favouring liberalisation has been sustained since the 1980s – has been the exception rather than the rule.[31] China, faced with bumper crops and falling prices in the late 1990s, ended open trade and tried to enforce floor prices using its public marketing agencies. This failed and policy reverted in 2001 to encouraging grain market development. India has put off liberalisation of grain marketing, at the expense of market development, despite widespread liberalisation in other sectors.

Third, privatisation of going concerns appears more viable for non-grain agencies, since they are more likely to have an established commercial operation independent of government. For example, privatisation of milk and cotton marketing agencies in Zimbabwe has been relatively successful, reflecting similar experience elsewhere (e.g. dairy privatisation in the UK). Privatisation of the cocoa buying agency (Produce Buying Company) in Ghana is proceeding. By contrast, privatisation of grain marketing agencies has not occurred in the case study countries. Efforts to commercialise NCPB in Kenya and GMB in Zimbabwe have been frustrated by political interference. Sri Lanka's PMB and Ghana's GFDC have remained non-operational for several years. Neither appears likely to be privatised as a viable business.

10
Can Public Services to Marketing be Contracted Out?

Public services to agriculture are in a state of change. The force for change is the ongoing effort, reflected in the 'new public management', to bring market relations into public services provision as much as possible. Contracting between buyers and sellers of services is one of the key components of this approach. This chapter first overviews contracting in relation to the new public management and the circumstances under which contracting succeeds or fails. Case studies of contracting out public services to agricultural marketing in developing countries are then presented and discussed: food security rice stocks in Sri Lanka; wheat milling in Sri Lanka; public grain storage in India; exporting maize from Kenya; and rural feeder road construction and maintenance in Ghana. Lessons from the case studies are brought together in the conclusion. The main observation is that managing reform of public organisations, rather than managing contracting out, is the underlying constraint on making NPM-style public management innovations work.

The new public management and contracting

On the liberalisation agenda, deregulation and privatisation are the main vehicles for the state to improve private goods provision, while corporatisation and contracting are the main means for bringing market principles into public spending and buying decisions. Corporatisation sets the department or unit up as a separate cost centre, or, in the case of a public utility (e.g. electricity, port authority) as an autonomous firm with a board of directors appointed by the state.

Contracting is central to the vision of the 'evaluative state' since it has the potential to build in competition, facilitate specification and monitoring of policy objectives, strengthen management accounting systems, strengthen quality control in service provision, and to establish a basis for consumer rights to a specified level of service (Walsh 1995: 112, citing Mather 1989).

Within NPM thinking, contracting is applied from the level of the individual public service manager, employee and department (performance contracts), to exchanges with other government departments, and with private firms (management contracting, leases, concessions and contracting out of specific tasks). The principle is as far as possible to subject all resource use decisions to actual or potential competition.

Contracts may be large and long-term (e.g. a thirty-year concession awarded for water supply to a French city), or they may be relatively small and short term, as in the case of a catering contract offered to a local firm by a local council. Private firms are invited to tender for the contract to carry out the task, possibly against an existing government department.

Contracting out has long been established practice in most governments for major engineering works (buildings, roads, boreholes, tunnels), and for highly specialised services (e.g. consultancies on policy design and impact). What is new is the greatly increased extent of contracting out on a routine basis, both of incidental tasks, such as cleaning, catering, security and maintenance of buildings, and of substantive departmental activities, where both of these were formerly carried out by government employees.

Contracting out of substantive tasks by government is used within a variety of agricultural services, such as storage, transport, procurement of agricultural inputs and their distribution, veterinary laboratory services, vaccination campaigns, information dissemination and small farmer credit schemes contracted out to banks. Longer-term management contracts or concessions (i.e. where the contractor charges users and is responsible for maintenance and investment on own account) are commonly awarded for running local market places.

When contracting out succeeds or fails

The underlying theory of contracting out derives from Coase (1937), who argued that in a world of well-developed firms and minimal transactions costs (i.e. where the costs of arranging, monitoring and enforcing a contract are very low), vertical integration of firms will be minimised. Firms would internalise only the costs of their main activity; linked activities upstream (e.g. inputs supply), downstream (e.g. marketing), or incidental to carrying out the main function of the firm (e.g. financial management, cleaning) would be carried out by other firms. Conversely, where vertical integration is in practice common, we expect that transaction costs are high and/or that other firms are generally inadequate to carry out these linked and incidental tasks.

The incentives within the public sector affecting the decision whether to contract or to integrate vertically, are different from those facing a private firm. Although the cost advantages for government of contracting may potentially be greater than those for private firms,[1] governments – even those facing extreme budget cuts – often lag behind private firms in contracting, because they are constrained by rules, by civil servants' conditions of service and by lack of a direct profit incentive.

The preconditions for successful contracting out in the public sector include:

- the private sector is active, with good firms competing for contracts
- government has the resources, experience and will to manage competitive tendering and contract monitoring
- the task can be clearly defined
- a 'contract culture' is well established, so that government is used to offering contracts and firms are used to tendering for them, and government is used to monitoring, measuring and evaluating performance.

The main problems encountered with contracting out include:

- specifying the tasks to be carried out under the contract. This becomes more difficult the more complex is the task, and the 'correcting mechanism' to be included, in order to fulfil social objectives. For example, carrying out health-care work is complex, since the objectives cannot be precisely defined and the appropriate methods to be used will vary with the disease, condition of the patient etc.
- lack of experience and capacity in monitoring and enforcing contracts, particularly when the task is complex
- fluctuating, unpredictable costs and revenues from the task. These make for high risk and uncertainty, which will need to be spread between client and contractor, and incur additional monitoring and negotiation costs. Clients who over-specify the details of a contract in the face of risk, are implicitly unwilling to share the risk; this is often sef-defeating since contractors will include a risk premium in their prices (Walsh 1995:124)

- inflexible civil service pay systems, not allowing individual performance-related pay. This is thought to be the main reason why performance contracting in government has worked relatively poorly in Africa (World Bank 1994b: 43)
- unaffordable subsidies, e.g. on consumer prices of grain and on farm inputs. For example, the strategy of commercialising agricultural marketing boards by separating their 'commercial functions' from the 'social functions' they carry out on behalf of government, has been hampered by inability of government to pay the cost of the 'social functions' which frequently consist of subsidies
- time-consuming tendering procedures
- collusion among potential contractors.

The ideal contract is a discrete exchange between independent parties. However, when the task to be undertaken is complex, there is uncertainty and risk. Continuous contact, trust and mutual understanding are then required between contractor and client in order to solve problems as they arise. Thus the promise that contracting out seems to hold of substituting market relationships for hierarchical structures is reduced as a task becomes more complex, involving 'the re-creation of management systems that are essentially hierachical in character' (Walsh 1995: 136). Indeed, '[t]he further one moves from the model of a discrete exchange between private actors, the more abstract the notion of a contract becomes' (Harden 1992: 36, quoted by Walsh 1995: 137). The complex contract ('relational contract'), requiring a degree of sharing of risk and difficult decisions over time between contractor and client, moves the relationship in the direction of vertical integration, and away from the idealised exchange.

Contracting out a service together with capital investment creates a complex contract, since it requires an ongoing relationship between contractor and principal in order to deal with unexpected events. BOT (build-operate-transfer) contracts are an example. As with a services contract, a complex contract should be specified as far as possible in order to confine risk and uncertainty. Risks should be carried as far as possible by the party best able to bear them (e.g. political and policy risks by government, production and trading risks by the contracting firm).

Sri Lanka: contracting out to reduce costs of food security[2]

The activities of the Food Commissioner's Department (FCD) are examined as examples of 'contracting out' arrangements between public and private sector organisations. The cases which follow highlight how the commercial self-interest of private sector agents can be exploited by government to reduce the financial costs of carrying out food security functions under a controlled food trade regime.

The Food Commissioner's Department (FCD), established in 1941, has a long history in the context of the food sector in Sri Lanka. The FCD was originally charged with responsibility for the importation, storage and distribution of the three major food commodities: rice, sugar and wheat. Following the privatisation of sugar imports in 1987 and progressive changes in the arrangements governing rice imports between 1988 and 1994, the activities of the FCD have been reduced to handling the distribution of wheat flour.

In addition to its operational activities, the FCD has a key role in the food security of the country. The Food Commissioner chairs the weekly Food Security Committee which comprises representatives from all departments involved with food production, imports and distribution and which meets to consider the prevailing food situation in the country in all its dimensions, notably, production, prices and imports. The FCD is also charged with the maintenance of the food security stock of the country.

Contracting out food security rice stocks

The FCD was the sole importer of rice into Sri Lanka until 1988 when limited private sector imports were permitted subject to licensing and quota restrictions. In 1994 all quota restrictions on rice imports were lifted and private imports are now permitted subject to duty being paid.

In order to provide a strategic reserve of rice within this liberalised market, in 1990 a system of bonded warehouses was established for rice importation. Under this system, contracted companies are authorised to import and store rice without paying duty until the stocks are released into the domestic market. No duty is payable if the stocks are re-exported. In exchange for this concession, bondsmen agree to maintain a minimum stocking level, varying between 1000 and 5000 tons per company. The stocks, which must meet specified quality standards, are held in FCD-owned warehouses for which storage fees are paid. Authorisation must be received from the Food Commissioner to release or re-export stocks which cause these minimum stocking levels to be breached. The bondsmen are free to import any quantity of rice but the FCD sets a minimum price at which stocks can be released into the domestic market.

The original contract was limited to three companies which shared a maximum quota of 200 000 tons of rice, allocated among them in lots of 100 000, 60 000, and 40 000 tons. Since 1990, new companies have joined the scheme, rising by the late 1990s to eleven licensed bondsmen able to import rice without paying duty. Only one of the companies was domestically owned. This contractual arrangement between the private 'bondsmen' and the FCD permits the maintenance of a buffer stock of one month's consumption of rice without any significant budgetary costs being incurred by the FCD. The contract requires that all handling and storage expenses are financed by the companies which also take responsibility for stock control and rotation. Not only does the system reduce budgetary outlays but, because stocks are held in FCD warehouses, the FCD gains revenue from the arrangements.

If a food security stock is to meet its purpose, adequate stocks must be maintained and readily available in the event of the need to draw them down. In addition to the quantity considerations, stocks must be of the requisite quality. If these conditions are to be met, either operators must have organisationally specific incentives which are consistent with their achievement, or a monitoring and enforcement system must be in place.

The contractual arrangement between the bondsmen and the FCD performs well when judged against the above criteria. The FCD has established that a food security stock of 50 000 tonnes of rice is required for Sri Lanka. This is equivalent to approximately one month's consumption. As an island economy with well-developed shipping facilities the stocking level is sufficient to meet domestic requirements during the time lag between an import order being placed and the arrival of the goods.

In terms of ensuring appropriate behaviour on the part of the operators, the contract between the bondsmen and the FCD relies on both private incentives and a monitoring system. As private companies the objectives of the bondsmen revolve around the generation of profits from their operations. In order to be profitable the operations must cover transaction costs (the maintenance of an import office, for example), the costs of storage and port charges. In addition to exempting their imports from duty until they are released on to the domestic market, the FCD can vary the fixed price below which released stocks must not be sold. By varying these conditions the FCD can alter the profitability of rice imports allowing it some leverage over the behaviour of the bondsmen.

A monitoring system is also in place which allows the FCD to ensure that the bondsmen comply with their contractual obligations. The stocks are held in warehouses belonging to the FCD facilitating both the monitoring system and access to stocks in the event of them being required by the government. Under current arrangements, stock data are compiled on a weekly basis and the information is faxed to the FCD in time for processing for the weekly meeting of the Food Security Committee.[3]

The system appears to be sufficiently flexible to balance the incentives of the bondsmen with the needs for a food reserve. Thus the Food Commissioner authorised the re-export of reserve stocks in 1995 when it became apparent that the bumper domestic paddy harvests and the wheat subsidy which reduced rice demand had made rice imports unprofitable. By exercising discretion in this area the Commissioner ensured the ongoing interest of the bondsmen in the arrangement.

But this raises the question of the sustainability of this contractual arrangement over the long term. Sustainability depends on the continued presence of private incentives for the bondsmen, the monitoring capacity of the FCD, and the capacity of the FCD to continue to structure incentives for the bondsmen in response to changes in the domestic and international rice markets. Sri Lanka is nearly self-sufficient in rice production; if import requirements fall below a threshold level private imports will prove unprofitable and the arrangement will falter. Between the years 1990 and 1994 incentives were such that the bondsmen freely maintained stocks above their minimum requirements. 1995 was an unusual year which nevertheless served to highlight the potential pitfalls of the arrangement given near self-sufficiency. In addition, the experience of that year demonstrates how important it is for the government to ensure that interventions are consistent with one another. For example, the impact of the wheat subsidy on demand for rice was a major contributory factor to the unprofitability of imports and could have jeopardised the viability of the bondsmen scheme. The raising of the rice import duty in the late 1990s to increase protection to local rice production reduced profits of rice import and in the long term constitutes a further threat to the bondsmen arrangement. However, it is arguable that a strategic grain reserve may no longer be needed once markets develop more, so that trade is more responsive. This is the position taken by South Africa, which has liberalised external grain trade without retaining a strategic reserve.

BOT contract for milling imported wheat

Wheat is an important staple food for Sri Lanka, second only to rice. Sri Lanka relies on imports of wheat, over which government still retains a monopoly for importing, milling and distribution. In order to reduce costs of milling, government in 1980 granted a twenty-five year monopoly wheat milling contract to PRIMA Ceylon. PRIMA is a private flour milling company headquartered in Singapore. PRIMA owns and operates an integrated milling complex at China Bay in Trincomalee, one of the world's best deep-water natural harbours. The contract combines a build-operate-transfer (BOT) arrangement and a milling agreement between the government and PRIMA Ceylon.

Under the terms of this contract, PRIMA constructed the mill and operates it. In the year 2005 the mill will be handed over in working order to the government at no charge. In addition PRIMA mills all Sri Lanka's imported wheat. The flour is returned to government. In exchange, PRIMA is guaranteed a monopoly of wheat flour milling and a minimum of 435 000 tons of wheat to be milled annually. From each 1000 kg of wheat it must extract 740 kg of flour with a quality specification of 10.5 per cent protein. Any flour extracted in excess of the 74 per cent and all by-products of the milling process are retained by PRIMA for private disposal. Most of the bran is pelleted and exported for animal feed.

Wheat is imported into Sri Lanka by the Coooperative Wholesale Establishment (CWE) according to a wheat purchasing plan, devised by the CWE on the basis of production and marketing of other food crops, specifying the quantity and timing of required shipments. This plan in turn is approved through a process involving the Ministry of Trade, Commerce and Food, the Treasury and the Cabinet.

Once the wheat has been milled, the FCD contracts private transporters (currently around 400) to collect the flour from the PRIMA site and move it to FCD stores and/or some 7000–8000 Multipurpose Cooperative Societies (MPCS) outlets. It has been argued that the distribution of flour supports the MPCS as they derive much of their income from its sale (Borsdorf 1993). The flour is then retailed to bakeries and the general public through both private and public channels. Legal ownership of the wheat and wheat flour is retained through all the stages by the CWE with payment for flour by the MPCS being made directly to the CWE.

The contract has been quite favourable to PRIMA. Edwards (1993) estimated the return on PRIMA's capital to be 20 per cent. PRIMA's milling operations are profitable at 435 000 tons of grain and yet it has consistently milled considerably more than this every year since the mid-1980s. PRIMA bears very little risk in the operation in the event of changes in international wheat prices as all procurement is financed by the Cooperative Wholesale Establishment. The major risk faced by PRIMA is that wheat grain will not arrive in a timely fashion meaning that the mill cannot be operated at a steady capacity. In addition, there is always the fear that the government will try to break the contract. But neither of these events has occurred.

The efficiency of the PRIMA contract from the government's stand-point can be assessed in terms of whether it is obtaining milled flour at a cost lower than would be possible if the milling facilities were government-owned and operated, or if flour were to be imported directly. The few studies which have been carried out suggest there are benefits to government. For example, Borsdorf 1993 (reported in Edwards 1993) argues that the costs of milling are not excessive in the context of milling operations more generally. This view was supported by a report by the Presidential Tariff Commission in 1993 which estimated a US$20/metric ton saving with domestic processing of grain (PTC 1993, reported in Edwards 1993), compared to importing flour. Furthermore, the transfer of the mill to government ownership in good working order in 2005 should be a substantial capital acquisition. In addition to cost savings, the wheat flour operations provide income to the Cooperative Wholesale Establishment (CWE), the country's largest state trading organization.[4]

Whether the PRIMA contract will be extended after 2005 is uncertain. As the end of PRIMA's contract approaches there is increasing debate over future arrangements and pressure from donors for the liberalisation of wheat trading. There has been pressure particularly for liberalisation of flour distribution, since stock losses by CWE have at times been excessive (e.g. 8 per cent during 1989–91).

In sum, the PRIMA contract demonstrates a working public–private partnership through a period when state monopolisation of wheat importing and distribution was the preferred policy in Sri Lanka. The milling contract has provided stable supplies, with returns to the milling company and to government. The possibility remains that liberalisation of trade in wheat and flour might compete down flour prices to the benefit of the Sri Lankan consumer. However, to date Sri Lanka has proceeded cautiously in liberalising its grain trade, keeping in place government controls and preferring partnership arrangements with a few private businesses to open competition, as demonstrated by the strategic rice reserve maintained by contracts with bondsmen, and by the PRIMA contract.

Kenya: contracting out maize exports[5]

Maize marketing in Kenya was formerly tightly controlled and confined to the National Cereals and Produce Board (NCPB). The Board had wide-ranging powers in support of food security objectives, through a combination of monopoly status in maize trading, administrative controls over maize movement and prices, and control over private sector maize trading activities. Large recurring losses by NCPB and rumours of corruption motivated reform.

In December 1993 the Kenyan government liberalised external trade in maize, allowing private firms to import maize and the National Cereals and Produce Board (NCBB) to export. This reform followed removal of barriers to private domestic trade in maize in the late 1980s. The reforms had been heavily donor-driven, but the positive response of business encouraged government to proceed further. In 1995 NCPB was required to put maize exports out to tender. Tendering for maize exports was a move away from the previous situation where the NCPB exported maize. This had incurred heavy costs and involved NCPB supervision of all the processes between collection of maize from the depots and discharge at the port of Mombasa. The purpose of introducing export contracting was to minimise losses from exports. This case study reviews the performance of the first two export contracts for maize under this new system.

Increased maize crops in 1994 and 1995 created the need to export. High transport costs for moving maize from the main growing areas in the west of the country to Mombasa have always prevented Kenyan maize exports from being profitable. Therefore maize export is only undertaken when there are surpluses to be disposed of. Even though maize prices in the world market in 1995 were at their highest levels for many years, they were below the into-depot price of the NCPB maize.

Contracting arrangements

Overall responsibility for maize exports was taken by the newly formed National Food Security Technical Committee (NFSTC). While the NCPB managed tendering, award and execution of the contracts, it was accountable to the NFSTC which monitored progress with the exports and reported weekly to Cabinet.

There were two exports of maize in 1995. The exported quantities were the stocks over and above the mandatory three million bags maintained as strategic reserves and privately held stocks estimated at ten million bags (NCPB Reports 1995; MOALDM Reports 1995).

The tenders for the first lot of two million bags were floated in June 1995. The offers were generally low, resulting in an outright financial loss of Kshs. 990 per bag. The tenders for the next export of four million bags of maize in 1995 were better designed. The government invited tenders for quantities which each contractor felt capable of handling at the prices they themselves quoted. The government did not give any indicative prices. This provided incentives for contractors to give the best prices possible and also to indicate what quantities of maize they were willing to export. Five major international commodity dealers were each awarded contracts for a negotiated quantity of maize.

Kenya's embassies in the tenderers' countries were used by the NCPB to provide confidential information on the tenderers. This is a role which the embassies had not played before. It proved useful because many of the prospective tenderers were found to have had poor business bases (NCPB Reports 1996). Even though the tender was an open one, the NCPB was accused of having influenced the selection such that the majority of those selected were the politically well-connected international commodity dealers in Kenya (*The Economic Review*, 2–8 January 1995, pp. 10–11). Interviews with top industry personalities lent credence to the allegation.[6]

The exporters were to obtain maize from NCPB depots in Mombasa and Kisumu as per agreed schedules. NCPB took the responsibility of ensuring that quality maize was moved to the two depots from depots in various parts of the country in large enough quantities to match the shipment schedules provided by the exporters. NCPB was also responsible for fumigation, organising the loading gangs and ensuring that the railways and road transporters adhered to their schedules of deliveries. The exporters had only to ensure inspection before loading on to ships and to provide the vessels as planned.

To deliver maize from various NCPB depots to the port in Mombasa, and to Kisumu for export to Tanzania, tender were invited for both rail and road transportation, to create greater competition in the bidding. The two are relatively well developed in Kenya and there is therefore considerable competition between them (Lewa 1995). Prices bid per tonne per kilometre, the reputation of the transporters and the condition of the trucks were taken into consideration in awarding contracts.

In order to ensure quality, inspection of grain by a private company was agreed between exporters and NCPB. The company, SGS, worked with the government's Chief Grader in Mombasa. The inspection costs were met by the NCPB. Inspection arrangement worked very well for the maize shipped through Mombasa. But there were problems of quality of the exports through Kisumu, particularly towards the end of the contract, since SGS did not certify grain quality in Kisumu. A further problem was that World Food Programme (WFP) which had contracted with the exporters for the Kenya maize in Tanzania, had higher quality requirements than those agreed between NCPB and the exporters. This caused much renegotiation, delay and rescheduling.

Performance

Overall, the experience with maize export contracting has shown that the capacity to handle contracts of such magnitude exists within Kenya. Contracting out resulted in a number of benefits:

- The contractual arrangements helped speed the flow of payments: NCPB was paid in advance for its maize by exporters, and of Kenya Railways and truckers, Kenya Ports Authority which was paid as per its contribution, Kenya Police personnel who supervised movement of maize were given allowances over and above their salaries, and farmers were paid more quickly than usual for their maize sales to NCPB.[7]
- Overall supervision by government through the National Food Security Technical Committee worked well.
- Costs of export appeared to be reduced, and there was reduced overall load on NCPB, despite NCPB's taking on the burden of managing the tendering and execution of the contracts.
- The tender process helped reduce delays (e.g. in processing documents at the port, in loading at the port, in moving maize from up-country to Mombasa by the Kenya railways) and improved quality control.
- Competitive tendering reduced transport time and costs in getting maize to Mombasa. It also greatly cut the time taken in loading maize on to the ships by the private operators.[8] This was due to their ability to pay overtime in order to quicken the pace of loading.
- Price risks were transferred to the private contractors. The NCPB negotiated for payments to be made in advance at the fixed agreed prices. NCPB therefore faced no risk of price fluctuations and exchange rate movements.
- Internal movement of maize generally worked well and without delays, with road and rail transport synchronized with the arrival of ships and the Kenya Ports Authority. The exception was shipping to Kisumu on Lake Victoria, which caused delays.[9]

Despite the overall success of the contracting-out of maize exports, a number of problems were encountered:

- Demands on resources: management of the tender process and contract implelementation was demanding for NCPB's senior management[10] and that of government ministries' officials in the National Food Security Technical Committee, who had the responsibility of monitoring and evaluating the progress of the contracts.
- Towards the end of the contracts coordination between maize supplying depots, Kisumu and NCPB headquarters became very poor[11].

- Failure to observe quality standards required for the Kisumu export contract resulted in much delay.
- Failure to meet export deadline: the original export contracts between the NCPB and the grain exporters were to expire at the end of March 1996. After the expiry, the contract allowed for a grace period of two months, and then further monthly extensions to June and September 1996 for those exporting maize through Kisumu. Government then refused to give any further extensions due to fears of imminent maize shortages. This put those who had prior commitments with importers in Tanzania and other countries in the region in difficulties.
- While the maize export contract was still being implemented, government introduced a 2 per cent presumptive tax on all exports of maize in June 1996, to discourage exports since maize shortages were imminent. This reflected the conflicting objectives of government in extending the export contracts.
- Private contractors complained of unpredictability in government policy regarding external trade in maize and wanted improvement of the Nairobi–Mombasa highway and rail services.

India: grain storage contracting[12]

At least since the Bengal famine of the early 1940s the Indian state has placed limited trust in the private sector, fearing collusion to raise prices rather than trusting in trade, competition and market development to stabilise prices. This has resulted in large-scale government intervention in grain purchase, storage and distribution by Food Corporation of India (FCI). In the grain storage sector this policy has recently been reversed. Rising costs of storing increasing quantities of grain, and high storage losses (Umali-Deininger and Deininger 2001: 329) have produced loud criticism. Policy now is to attract investors into storage in order to shift financing costs towards the private sector and to raise quality. Increased incentives are offered, including extended guarantees by FCI to pay for contracted storage for ten years and more, even if not used.[13]

This case study explores FCI's contractual relations with public and private providers of storage. It notes that contracting out of storage is a long standing practice in FCI, and that FCI has favoured public over private providers of storage. Furthermore, the continuing underdeveloped condition of the private grain storage market may be partly the result of inter-seasonal grain price stabilisation through FCI. In a developed grain market, private storage and trading in futures and options contracts play the critical role in inter-seasonal price stabilisation.

Current arrangements

The policy of the government of India has been to ensure national food security by maintaining large physical stocks of grain in the country. These stocks are obtained through FCI's procurement operations. The minimum size of buffer stocks to be maintained is fixed by the Ministry of Food and varies by season, off take and sales by FCI. For instance, stocks in April are low because procurement of wheat starts only after 15 April – the peak procurement season for wheat is April–July; similarly, the peak procurement season for rice is December.

The large scale of buffer stock operations has required creation of stores at strategic locations. FCI is the main agency which provides storage for food grains. In addition to constructing its own go-downs, FCI also hires storage capacity from other public agencies and private parties.

It is necessary for the FCI to hire storage capacity because FCI-owned storage depots are insufficient in relation to the scale of procurement and more than half of them are concentrated in the northern zone (i.e. the surplus food grain states). Hiring of storage capacity is done from several agencies – public as well as private on a spot or lease basis.[14] Among the public agencies are the Central Warehousing Corporation (CWC), state governments and the sixteen State Warehousing Corporations (SWCs). Storage from both private and public agencies is obtained on the basis of identified needs and requirements by each zonal/regional office of FCI for their respective regions.

Performance of the arrangements

FCI appears to have no difficulty in arranging storage, although during the procurement season there may be shortages in particular regions. However, much of the storage is of low standard,[15] and guarantees and financial inducements have been needed by both public agencies and the private sector to encourage investment in grain stores, particularly large stores.[16] Guarantees have been given by the FCI to both CWC and private parties prior to construction of grain stores, in the form of long-lease contracts renting the store, whether it is used or not. Financial assistance for constructing stores was originally channelled through the Agriculture Refinance and Development Corporation.[17]

FCI prefers to hire from CWC and SWCs where possible. Private storage is sought only if no storage is available with public agencies. The reasons cited for this preference by FCI officials were greater trust in the public sector, close long-term association with the CWC (the FCI Board of Directors includes representatives of CWC), and greater reliability and certainty in terms of availability as well as expected quality. FCI virtually guarantees use of about half of CMC's storage capacity. By regulation too, the CWC is expected to earmark at least 40 per cent of its storage capacity for food grains storage under the national food security policy.

FCI staff claim that superior quality is the reason for their preferring to hire public sector grain stores:

- Size of the go-downs: the sizes of CWC go-downs range from 5000–50 000t capacity whereas the average size for private go-downs is estimated to lie in the 1000–2000t capacity range.
- Provision of railway sidings: private go-downs are less likely to have these than CWC godowns.
- Weigh bridges, handling and bagging equipment etc.: CWC godowns are better equipped.
- Location: it is more difficult for a private go-down to be located adjacent to a national highway/railhead than a public agency like FCI or CWC, for the simple reason that land acquisition in such areas is relatively easier for a public body.
- Maintenance: preservation of quality of food grain stocks which involves continuous prophylactic and curative treatments is reported to be better at CWC warehouses, which have more advanced storage and preservation techniques.

With regard to cost, hiring from the private sector appears to be cheaper, although FCI staff claimed there there were no cost differences between hiring from CWC or the private sector. Prices paid to private parties in the mid-1990s were 144 paise per square foot while those to the CWC were 140 paise per bag. As reported by FCI staff, one tonne of foodgrains requires six square feet of storage space and 10 bags equal one tonne. Based upon this information, hiring charges paid by FCI to CWC would be 1400 paise per tonne and to private agents would be 864 paise per tonne.

When cross-checked with CWC and private owners, it was confirmed that CWC charges were higher than those of the private storage owners – the difference (according to CWC) is attributable to better quality of maintenance and more reliable security systems. Associated with attributes of lower risk and uncertainty are higher operational and administrative costs.

Incentives to the private sector vary from region to region and are obviously demand-related. Interviews with some private operators around New Delhi indicate that initial capital investment in construction of a storage facility of 5000t capacity is very high, finance for which is difficult to arrange. Closely related to high investment costs is the degree of risk – uncertain business conditions with respect to demand for the facilities. Those individuals who benefited from the ARDC scheme and have hired to the FCI in the past, reported that their construction costs were amortised within 5–6 years. The latter were unequivocal that if finance had not been arranged and business assured, they would not have incurred such a big risk and ventured into this enterprise.

The situation with respect to smaller storage facilities of, say, 1000t capacity, is somewhat different. Business activity here is more visible, mainly because capital costs are lower. Further, hiring out for this group is profitable as it is a related/additional business rather than their sole activity. This is especially true for non-surplus states where the scale of food grain trade is low. The opportunity costs of keeping the facility empty being high, it is attractive for them to hire it out on FCI's terms and conditions.

In conclusion, the coexistence of long-established storage contracting with a less-than-buoyant private storage market might be the result of limited inter-seasonal grain price variation, caused by FCI's price stabilisation operations. While these operations require much hired storage, this is supplied substantially by state-owned storage oganisations (CWC and the SWCs). The return to private storage is insufficient to stimulate private investment in dedicated grain stores. In a developed grain market private storage, allied with futures and options dealing, plays the critical role in inter-seasonal price stabilisation. For this role to be entrusted to the private sector (as South Africa has recently done, see Chapter 8 above) requires much confidence by the state in the benefits of private trade. It also requires confidence in government policy by private investors.

Recent policy change favouring private investment in storage capacity is contending with subsidised public facilities and institutional arrangements which favour FCI's use of CWC and state corporations. Therefore the entry of larger-scale private investment into grain storage might be through joint ventures with CWC and other state corporations. Recent freeing of CWC to increase its scope and invest abroad, plus increasing joint ventures of Indian companies with private investors, particularly in container facilities at ports,[18] suggests government is trying to further commercialisation of parastatals in this sector.

Under present circumstances India's hopes for a private sector solution to storage for a substantially unreformed PDS, by hiring from private investors who construct and own good quality facilities, may require more confidence by investors in future policy on grain distribution than exists. This may require a clearly articulated and comprehensive reform strategy, which India has so far been unwilling to put in place.

Ghana: labour-intensive contracting for feeder road rehabilitation and maintenance[19]

In cocoa exporting countries, state investment in roads tends to follow booms in cocoa production (Ruf 1995: 80–1). In Ghana supply of cocoa roads has typically lagged well behind demand. There are two government organisations which are responsible for feeder roads – the Department of Feeder Roads (DFR), currently part of the Ministry of Highways, and the Feeder Roads Department of Cocobod, which looks after the cocoa areas. A case study of contracting by DFR in order to construct and improve feeder roads is presented here.

Background

There are large price differentials between producer and consumer areas across most of Ghana. It is commonly thought that a large proportion of the difference is due to the expense of transporting the goods. This expense is composed of a number of items: wear and tear on vehicles and the cost of spare parts; fuel and other recurrent costs; and vehicle purchase price. Government has an indirect influence over all of these costs. Fuel is taxed heavily and typically constitutes over half the cost of transport; parts and vehicles are scarce and expensive in an economy where low demand limits economies of scale for traders.

Ghana is often described as a 'footpath economy' (World Bank 1993). The main roads are generally passable – with some notable exceptions especially in the remoter and northern regions. The reliance of large numbers of producers on footpaths to get their crops to a road head disadvantages them compared to farmers with access to a feeder road. In a study carried out in the Kumasi area 'the improvement of existing road surfaces was estimated to have a negligible impact on prices paid to the farmer. However, connecting a village to a road head by converting a footpath to a vehicle track was calculated to have a gross beneficial effect in the order of a hundred times greater than improving the same distance of earth track to good gravel road' (Hine and Riverson 1982: 85). The same study concluded that, as far as agriculture was concerned, the quality of the road was of minor importance. Scarce engineering resources were best used in keeping roads open – maintaining bridges, culverts and other small-scale remedial work (ibid.: 89).

Given this situation, one would expect a substantial degree of public investment in transport systems, in particular road infrastructure. Despite the high returns to feeder road investment, they appear have suffered the highest degree of neglect, while the main roads have received considerable investment and maintenance.

Labour-intensive feeder road rehabilitation and maintenance

Ghana has a large network (21 000 km) of feeder roads which had fallen into disrepair by the early 1990s. Only 3 300 km were usable during the rains. DFR is charged with rehabilitating feeder roads which will then be maintained. The objective in the mid-1990s was to get 60 per cent of the network to a maintainable condition.

The government of Ghana has always built and maintained roads through private contractors. The innovation in the feeder roads programme is to introduce the idea of 'labour-based' contractors. These rely on labour rather than machinery to do most of the work. DFR was training the contractors and their supervisors. The objective was to have at least one such contractor in each of the 110 districts. By the mid-1990s sixteen contractors had been trained and equipped, and another twenty-seven trained but not yet equipped. These contractors were well placed to work on the scattered spot improvement works which are needed to bring roads to maintainable condition. Their survival is therefore important. They are cheaper than equipment-based contractors, even without an open tendering system. Tendering had not yet been introduced as the contractors were judged not ready for it.

But the progress in rehabilitating feeder roads is slow for a variety of reasons, some of them related to contracting arrangements:

- Although a number of donors (led by the World Bank, with Danida and USAID) formed a consortium to assist government to rehabilitate this network, the DFR had been unable to spend available aid money as rapidly as projected. The investment was $5 million per year in 1993; at this rate it would take thirty years to rehabilitate the feeder road network.
- The slowness may have been partly because the procedures for design of works were complex, requiring specialised consultants to go from Accra to do the design work. This took two months in the region and three months in Accra. Tendering then took one month, and a further two months were needed before the contract is awarded. The reports prepared by consultants all had to be screened and approved, leading to a huge amount of reading for DFR staff. As the number of documents prepared increases, the delays are likely to get longer.

- The capacity to carry out the rehabilitation programme is limited. The World Bank commented that the Department of Feeder Roads was understaffed, especially at middle management and technician levels. It was difficult to attract experienced engineers, and the younger engineers were only retained in government because training (especially overseas) opportunities were better than in the private sector.
- In many cases roads were over-designed, according to Danida's consultants. Donors have put pressure on the Department of Feeder Roads to reduce road standards – especially width – and go for spot treatment rather than full rehabilitation. However, the Department was reluctant to go very far in this direction as it believes that once roads are rehabilitated, traffic, especially heavy traffic (tractors and lorries), increases. There was no traffic information available for most feeder roads to enable appropriate technical solutions to be found.
- Progress with evolving a policy on maintenance and a 'maintenance culture' was also slow.
- Contractors complained of the fixed rates paid by Department of Feeder Roads, its slow payment procedures, lack of payment for reinstating rain-damaged work, and problems repaying equipment loans to banks.
- In the regions the capacity of the contractors could be low. Few contractors in the northern region had competent supervisors. The DFR resident engineer was still needed to offer technical supervision. Contractors were unable to undertake design work.
- Lack of capacity in aid agencies was a further source of delay, causing delays in disbursement of IDA and OPEC funds.

In sum, although the capacity of the DFR was being built up to meet the challenge of feeder road rehabilitation and maintenance, it was a slow process, and very dependent on continued donor funding over a long period of time, as well as the allocation of greatly increased government budgets for maintenance.

The options for speeding up the feeder road rehabilitation programme are quite limited. One proposal has been to move DFR into the Ministry of Local Government and Rural Development, and to strengthen the links with local governments around the country. This might enable some cost-sharing to take place, at least in the wealthier and better organised areas. The design of the programme envisaged that local government would eventually participate. There is already a pilot decentralisation programme, and there are District and Regional Tender

Boards for the issuing of contracts. District Assemblies are now involved in deciding priorities for the DFR's programme of work. Some DAs have started to use part of their Common Fund allocation from central government to contribute to the DFR's programme.

More radically, there is a need for greater cost recovery, to generate resources for maintenance and construction. This may be done through toll systems, or where there are stable sets of users, by encouraging them to associate for purposes of investment in feeder roads – for example, in cocoa and other export or industrial crop areas as stable marketing arrangements evolve. It may be easier to do this for feeder roads than main roads, where there has been little uptake by the private sector of the government's build-operate-transfer scheme.

Conclusion

The NPM literature stresses the need for capacity in government to manage the contracting process, leading to the expectation that this may be the critical constraint on making contracting out work well in developing countries. The case study of contracting out of feeder road construction and maintenance in Ghana does point to capacity constraints on the process, particularly on managing the technical aspects. But in the Indian, Kenyan and Sri Lankan case studies managing the contracting process is not the constraint (see Table 10.1). These studies also indicate that contracting out by government departments in developing countries is not a new practice. There may not always be a level playing field among tenderers (e.g. in India, FCI's favouring of CWC despite higher costs) but contracting out is familiar and much used.

In short, ability to contract out may not be the main constraint on adopting competitive contracting in public management in developing countries, and on securing benefits from it. Rather it is the ability to carry out needed reforms in organisational arrangements which is frequently the constraint. This is suggested by two observations.

First, even where competitive contracting has been adopted widely for service *inputs* into government it has not been adopted widely for service *outputs* of government. In Zimbabwe, where the Public Service Commission in the late 1990s demanded that ministries commercialise and outsource services wherever possible, services to public bodies (e.g. cleaning, laundry, security) were widely contracted out. But there was only one attempt to use competitive contracting as the reform model for a public service output to agriculture. This was in the Tsetse Control Unit in Zimbabwe's Ministry of Agriculture and Lands in 1998.

Table 10.1 Overview of case studies of contracting out

Country	Client	Task	Contractor	Impact	Associated reforms
Sri Lanka	Cooperative Welfare Establishment (CWE)	Wheat milling	PRIMA Ceylon Ltd.	Wheat milling monopoly in return for free milling of CWE's wheat imports and BOT contract on mill. May raise costs to consumer by limiting competition	None
	Food Commissioner's Dept	Strategic reserve of rice	Bonded traders	Shifts costs of strategic reserve on to private trade, in return for import licence. May raise costs to consumer by limiting competition	Switch from universal rice ration to targeted rations, and decline of Paddy Marketing Board (PMB)
Kenya	National Cereals and Produce Board (NCPB)	Export of maize	International grain traders	Substitutes contracting by NCPB for direct export by NCPB. Unclear whether there were savings to NCPB	None
India	Food Corporation of India (FCI)	Storing rice and wheat	Central warehousing Corporation (CWC), state warehousing corporations (SWCs), private grain storage companies	CWC costs exceed private costs. FCI need for excess storage results from unreformed public Distribution system (PDS)	None
Ghana	Department of Feeder Roads (DFR)	Feeder road construction and maintenance	Labour-intensive contractors	DFR rigidities and resources limit effectiveness	None

Putting this into action was stalled by lack of capacity in the unit to transform itself from an implementer to a regulator, and complicated by the desire of the former director of the unit to compete for the contract.

Second, promotion of contracting out may contribute little substantial benefit, even where it is efficiently carried out, if it takes place in the context of wasteful, unreformed organisational arrangements. Contracting for grain storage by FCI in India and for maize export by NCPB in Kenya is routinely managed. But any resulting benefits in reduced costs or increased quality have been dwarfed by the underlying inefficiencies in the policies these organisations implement – policies which generate the need for excess public grain storage in India or public exports of maize from Kenya. By contrast, the contractual arrangement with bonded rice importers in Sri Lanka to hold minimum stocks of rice followed abolition of the universal rice ration and its substitution by a targeted ration. This reduced public sector buying, selling and stocking of rice by the PMB, and suggested the need for an alternative strategic grain reserve. Among the case studies of contracting out (see Table 10.1) this stands out as having the clearest benefits. Significantly, the contracting authority in this case is a steering and facilitating body (the Food Commissioner's Department) not a service-providing organisation, which therefore did not need itself to be transformed in the process, but was able to manage the reform.

11
How can Quality be Assured?

Traditionally the public role in quality assurance for agricultural produce has been considered high compared to other goods because:

- *external effects of agricultural produce and food processing can be great*: the public is especially concerned about diseases (e.g. salmonella, e-coli, BSE) and cancer-related substances (carcinogens) carried in food and feed. These require standards, enforced by inspection and sometimes by licensing in the case of premises (food processing, restaurants). Public regulation against infectious diseases (e.g. in water, animals, plants, seeds, feeds, food) is present in most countries, with enforcement by government agencies, central or local, and varying greatly in effectiveness.
- *reputation effects are limited for agricultural produce*: agricultural produce is in many cases a commodity produced by numerous farms, rather than a unique product of a recognised firm. Reputation effects are stronger where the product is clearly identified with the seller (branded goods, goods immediately consumed, e.g. restaurant food) and where quality attributes are visible ('transparency'). Neither has applied strongly to unprocessed agricultural produce.

The expansion of the role of the state in the mid-twentieth century included quality assurance bureaucracies, sometimes restricting trade and private ownership with this purpose (e.g. seed production and distribution in Zimbabwe). Public roles were to screen, set standards, facilitate and promote industry brands of quality (often amounting to country brands of quality, as in the case of the market premium earned by Ghanaian cocoa).

However, current food market development in high income countries is strengthening reputation effects, as a result of supermarket dominance and raised concern among consumers about food safety, food production conditions and environmental impact. The dominance of supermarkets is leading to more exacting standards, as they drive and implement quality control right down the marketing chain, including food imports. With consumers demanding traceability, plus more characterised and differentiated products, there is the opportunity for producers to brand their formerly undifferentiated produce (estate wines and coffees, fair trade, organic) and thereby in turn drive quality up the marketing chain.

All this leads to a shift in need for government intervention, away from routine grading and inspection (which is done within the industry, particularly for exports, to the standards set by the buyers) towards a stronger regulatory role focusing on setting and monitoring public health standards (e.g. pesticide residues, meat and fish hygiene). Carrying out this role may involve: changes in central vs. decentralised responsibility within government (e.g. the recentralisation of meat hygiene services in the UK after the BSE and e-coli outbreaks); letting firms have greater routine responsibility for quality in line with the greater ownership of quality control within the production and marketing chain; and setting standards (e.g. the HACCP initiative[1]) which allow more flexibility to firms in managing food safety in their business.

This trend of change in markets and state intervention is strong in high income countries but weaker in lower income countries except in their export sectors. However, the pressures are now growing from nascent consumer movements in developing countries for safer food in domestic markets.

Our case studies are all of performance and adaptation in organisational arrangements which were set up in the mid-twentieth century to grade, inspect – and in some cases to market – agricultural products for export or domestic consumption. Where public monopolies were the means of quality assurance there is now a need and opportunity for reforming the public role, as a result of market development after trade liberalisation, and public demands for increased food safety.

India: the changing role of the Directorate of Marketing and Inspection (DMI)[2]

The Directorate of Marketing and Inspection (DMI) is an attached office of the Ministry of Rural Areas and Employment (MREA). Originally established in 1939 as the Office of the Agricultural Marketing Advisor to the Government of India it was intended to act as the central agency for initiating and coordinating state activities related to the development of agricultural marketing. Its role now includes implementation of agricultural marketing policies and programmes at the centre and the guidance/coordination of similar activities in the states.

The activities of the DMI cover a number of areas, one of which is to promote grading and standardisation of agricultural and allied commodities under the Agricultural Produce Grading and Marketing Act (APGMA) 1937. The promotion of grading and quality control by the DMI has three main objectives:

(a) protect the producer from exploitation;
(b) provide a means of describing the quality of the commodity to be purchased/sold abroad by buyers/sellers all over the country;
(c) protect the consumer by ensuring quality of the products purchased.

The DMI is legally empowered to enact rules for prescribing grade standards for agricultural and livestock products, defining quality, prescribing methods of marking, packing, sealing and laying down conditions for issue of 'certificates of authorisation' for carrying out grading under the APGMA, 1937. These specifications are popularly known as AGMARK standards: the name signifies agricultural marketing and stands for a quality seal ensuring quality/purity of the product. DMI's grading focuses mainly on fruits, food grains and finished products like spices, wheat flour and honey. These are mainly items of domestic consumption, export of which has started only in the recent past.

Corresponding to the above objectives, the policy on domestic quality control focuses on producers and consumers and is voluntary in nature, while the policy towards export quality control is compulsory. The approach towards domestic producers and consumers as well as exports is discussed below.

(i) Grading at traders' level: this grading is voluntary and initiated at the request of the manufacturer/packer. The target here is the consumer and the standards apply to processed as well as unprocessed foods. These foods have been classified by the DMI into centralised and decentralised products. Centralised products require elaborate chemical testing and are directly tested by the DMI, for example, ghee, butter, edible oils, powdered spices and honey. Decentralised products, for example, wheat flour (*atta*), rice and pulses are those products for which grading and standardisation is relatively simple. This grading has been delegated to the respective marketing departments of the states, though they are monitored and supervised by the DMI.

(ii) Grading at producers' level: the aim of the DMI in grading at the producer's level is to establish a uniform trading language to specify quality standards to enable producers to achieve higher market prices for their products. Simple standards have been evolved by the DMI to classify agricultural products into different grades, e.g. for apples. Similar standards have been evolved for mangoes, bananas, oranges, onions and potatoes. These grades are implemented through established grading units in the regulated markets where agricultural products are subjected to simple tests and graded prior to sale.

(iii) Grading for exports: historically, the DMI's role in ensuring quality control for exports dates back to the pre-independence period (1942) when the export of hemp was prohibited unless graded, marked and stamped under the APGMA, 1937. With the onset of the structural adjustment programme in India in 1991, export quality control was substantially liberalised and exporters are now allowed self-certification for export purposes. Informal estimates provided by the DMI indicate that nearly 90 per cent of export products are self-certified.

The DMI's coverage of commodities has gone up over the years suggesting that producers are becoming more willing to recognise the desirability of a quality control system.

AGMARK

The APGMA, 1937 (as amended in 1986), empowers the DMI to fix quality standards known as AGMARK standards; to prescribe terms and conditions for using the seal of AGMARK; to institute prosecution in case of violation of any provision of the Act; and to redress consumers' grievances with respect to violations of the Act.

The criteria on which the DMI evolves standards for agricultural/food products are both internal and external. First, prevailing standards in the trade are taken into account. Second, the DMI's own basic minimum requirements are incorporated before standards for the product are set. According to the DMI, these standards are likely to vary among regions or between urban and rural consumption zones. An example here is rice, of which nearly 8000 different varieties are known and cultivated in different regions of India. With so many varieties, it is impossible to have one uniform standard, so a common and simple grading procedure has been developed for rice based upon the length–breadth ratios of the grain. Existing standards are reviewed occasionally every four–five years. A revision is normally initiated when there is a marked shift in consumers' preferences or change in processing technology of the product concerned.

The market inspections of AGMARK products are carried out by the DMI marketing inspectors. These inspectors are graduates in engineering, chemistry or veterinary science and are allocated products in conformity with their educational backgrounds. Inspection of market products is carried out on a routine basis with the inspectors drawing random samples from different retail or wholesale shops. These are then tested on the spot, or taken to laboratories for chemical analysis. The lot numbers that the product is expected to carry are then tallied with the analysis report received from the manufacturer/packer's chemists for discrepancies, after which the DMI initiates action. Lapses that are unintentional such as moisture losses, transportation spoilage and so on are condoned. But intentional violations usually result in temporary or permanent suspensions of certificates of authorisation.

Extension

The focus has been on educating farmers about the advantages of grading their produce before sale and generating consumer awareness about the availability and use of quality products under AGMARK and the dangers inherent in consumption of adulterated foodstuffs.

The dissemination of quality control awareness by the DMI is carried out by organising and participating in exhibitions, seminars, workshops, screening of films, cinema slides, supply of publicity material, etc. At the centre, the DMI maintains active collaboration with the Ministry of Information and Broadcasting and its media organisations like the Directorate of Advertising and Visual Publicity, All India Radio (AIR), Doordarshan (public television channel), and the Directorate of Field Publicity for its publicity programmes. The DMI publishes a quarterly journal called *Agricultural Marketing*, that focuses upon quality control and standards in agricultural products, amongst other aspects of agricultural marketing.

Other public quality control organisations

In addition to the DMI, there are other government/quasi-government quality control institutions in India concerned with food.[3] The Bureau of Indian Standards (BIS) was established in 1947 as a regulatory body with quality control as its sole activity. It is therefore more specialised and focused on this activity as compared to DMI, which has many other organisational objectives and activities. The BIS trademark is ISI; it has a network of supporting laboratories and offices in major state capitals and a large body of skilled staff. A major distinction between the BIS and DMI is that the former's range of quality control activities is more comprehensive; its standardisation applies to practically all products except medicines, whereas the DMI is restricted in scope to agricultural/semi-agricultural processed produce only. There is some overlap between the two, since both undertake quality control for agricultural processed products (e.g. gram flour or besan).

Performance of the DMI

DMI estimates that it grades one-half of total output of wheat flour, but has no estimate of the proportion which it grades of all agricultural products. Regional variations in coverage are very large; for instance, the proportion of graded wheat flour marketed in New Delhi is some 90 per cent, but is much lower in rural areas where consumers bring wheat directly to small mills for grinding or where loose flour is purchased directly from the mill itself.

The AGMARK label is widely advertised and used by producers and traders. Its importance is well understood by the producers; because there is a price differential between graded and ungraded goods, they have an incentive to get their products graded. There is also some preference given to these products by the government where purchases of AGMARK labelled products are given a small price premium.

But DMI admits that inspectors find many instances of violations, misuse and abuse of grading standards. Issuing of show-cause notices or temporary suspensions by the DMI are quite frequent for voluntarily graded products. In contrast, quality control for products that have to be compulsorily graded – fat spread, blended oils, ghee, for example – is well established as penalties for violations are severe and extend to imprisonment (under the Prevention of Food Adulteration Act).

These repeated instances of products carrying quality control labels and yet not conforming to the required standards lowers consumer trust, especially in the case of spices. However, violations of this kind are not restricted to AGMARK products alone, as quality control is not yet taken seriously by most producers except insofar as graded products gain a price premium. BIS officials admit that they face the same difficulties as the DMI in enforcing standards.

An underlying reason in DMI's case may be that DMI's monitoring machinery for inspection of quality control does not seem to very effective. Given the numerical strength of its staff (fifteen marketing officers to cover a large city like New Delhi), it is doubtful that the necessary monitoring levels can be achieved as spot checks are rare for any single market.

Regarding financial performance, DMI covers some 80 per cent of its costs from revenues earned from selling grading services to producers and traders, with the balance covered by the government of India.

In domestic markets, the DMI would appear to be competing with other agencies like the BIS in the same products. In urban areas, the labels AGMARK and ISI are well known, the latter somewhat more so being applicable to a wider range of products. This is not true of rural areas though, where the DMI has yet to penetrate and the AGMARK label is not so widely known.

In export markets, DMI's role in export quality control is now virtually non-existent, mainly because of the abolition of compulsory quality control for exports, but also because other quality marks are often preferred (e.g. ISI and ISO 9000).[4]

Ghana: quality control for cocoa exports[5]

Ghana's cocoa retains its reputation as the world's top quality cocoa. This is recognised in premia paid, and in the confidence buyers have had in Ghanaian cocoa traded by the Cocoa Marketing Company (CMC), a wholly-owned subsidiary of Cocobod. It is in no small measure a result of the work of the Quality Control Division (QCD) of Cocobod, which has a statutory responsibility to control the quality of cocoa exports, and until recently coffee and sheanut exports too. QCD's system of quality control for cocoa is first reviewed, followed by the emerging quality control issues and options in the liberalisation process which began with the introduction of private domestic buyers in 1991 and was extended in 1999 to 30 per cent of export sales.

QCD's system of quality control for cocoa

The long-standing system of quality control for Ghana'a cocoa is an intensive but simple one. It is based on the fact that Ghanaian cocoa farmers – predominantly smallholders – are well-versed in the production of good quality beans. Most cocoa farmers have been in the business for generations. There is a good buying infrastructure (2700 buying centres); they are generally not in a hurry to sell cocoa before all the on-farm processing which ensures quality has been completed. This compares with the situation in Côte d'Ivoire (see below) where arrangements favour larger farmers who do their own marketing, and where small farmers often rely on itinerant traders whose appearance to buy cocoa may occur before it has all been adequately fermented and dried. Because the farmer is unsure when the trader will reappear, the process is interrupted. In Ghana, buyers[6] purchase fermented and dried cocoa from the farmers. They invite the QCD to inspect the purchase, grade it and seal the sacks at their local collection centre. For this a fee is paid, determined by the Producer Price Review Committee.[7] Following this, if cocoa is stored in buyers' warehouses for any period of time it has to be fumigated, in order to preserve its quality. The QCD then carries out further inspections at the point of takeover from the LBC and in the port stores, where a final fumigation may be carried out. In each location there are QCD staff employed to check the produce. The sealing of bags means that cocoa can be traced back to source, and faulty gradings dealt with. There are two grades of good quality cocoa, for which the producer receives the declared producer price. Sub-standard cocoa beans are taken out of the export system, and the producer or LBC penalised by a reduction of 50 per cent of the value. The key to the success of the system lies in the checking at buying centres. This reinforces the farmer's desire to present a good quality fermented bean.

QCD's role in a liberalised market

Even after liberalisation of the internal cocoa market and part of the export market, the QCD's services remain compulsory for all buyers and exporters, mainly because of government's commitment to maintain the premium quality of beans shipped from Ghana. Government's intention is that QCD will in time be confined to an overall regulatory role for quality assurance. The reason for maintaining QCD's monopoly for the present is its concern that private exporters may not immediately be able to maintain the required quality, with consequent loss of market share by the country. This anxiety feeds on the experience of countries like Cameroon and Nigeria which liberalised exports, and of Côte d'Ivoire (see case study below) which has abolished its state quality control process.[8]

A number of issues regarding quality control have emerged since liberalisation began in 1991. There has been discussion for some time at the Cocobod of privatising the QCD, and significant steps have been taken to commercialise its operations. The QCD has become semi-autonomous and increasingly self-financing by charging buying companies for the services it provides. The QCD is confident that if privatised it could compete well with the only other quality control firm in Ghana – S.G.S. Ltd. It believes that foreign buyers have confidence in its judgement. As part of its preparation for privatisation, a number of new specialist graduate staff were recruited, district level staff retrained, and top management trained in financial control. However, despite exceeding its revenue targets QCD had to shed some 15 per cent of its staff in the mid-1990s. The number of inspections of each consignment went down from five to three. Buying companies are now asked to carry out more of the inspection procedures themselves.

The views regarding QCD performance are mixed. Economic analysis of the benefits to Ghana of QCD's work has been positive (LMC 1996a: 30). Nevertheless, it suggested that there was room to reduce the costs of quality control. Traders generally perceived the service offered by QCD as a useful one, in that it maintains their reputation for supplying a quality product. Their view is that a statutorily imposed quality control process is necessary to prevent exporters who are under pressure to fulfil contracts with overseas buyers from exporting lower quality produce to make up their consignments. However, an international commodity buying firm operating in Ghana was critical of the service, and its compulsory nature. Taking the view that the company was already quality conscious, and already making its own arrangements for quality control both in Ghana and by sending samples abroad, the QCD's interventions were seen as causing unnecessary delay. QCD's methods for quality control in coffee and sheanuts were also criticised.[9]

An end-user's view (Cadbury's in Birmingham) in the mid-1990s was that previously only cursory inspection of Ghanaian shipments was necessary. But Cadbury's had begun its own quality control process, worried about the degree of control being exercised over LBCs' purchases, and as a result of QCD adjusting its quality control procedures to make them more competitive. However, Cadbury's staff felt that it should not need to be the role of the end-user to check quality and therefore looked forward to a return to previous standards.

There were reports in the mid-1990s of LBCs attempting to minimise fumigation costs by evading the compulsory services of the QCD, complaining that it leads to delays in the evacuation of their cocoa to the Cocoa Marketing Company's takeover points at the ports. Some also complained QCD staff are corrupt and require bribes before they will grade or seal sacks of cocoa (Nyanteng 1995: paras 153–5). Poor bean quality resulting from such evasion led to further delays in crop takeover as sampling for quality revealed the inadequate fumigation. Part of the problem is caused by a mismatch of administrative boundaries between the LBCs and QCD. The latter allocates its staff to districts on the basis of production figures; the LBCs sometimes amalgamate their operations over two or three districts. Such matters can be resolved through better communication between LBCs and QCD, and greater flexibility on the part of QCD. These efficiencies might be encouraged by a competitive quality control system.

Greater efficiencies might also be achieved by devolving as much of the quality control function as possible to the LBCs themselves. Cashpro (an LBC) already has its own quality control staff, for example, and puts some effort into educating farmers on various aspects of quality, and has imposed sanctions on purchasing agents who buy sub-standard crop. This is a development that QCD favours since QCD sees its role as nurturing the LBCs, educating them on the importance of quality control and how to achieve it, thereby taking on an enabling as well as a regulatory function.

There is evidence of QCD responsiveness to LBCs' problems. When QCD still had responsibility for sheanut export quality control, some exporters lobbied to remove inland QCD inspections, and to concentrate all inspection at the ports. A Cocobod-industry committee was established to revise inspection policy. However, while the bigger companies were quite happy with the proposed change, smaller exporters were less happy as there is little market for rejected sheanuts in southern Ghana. QCD, anxious to rationalise its operations, persuaded the smaller companies to go along with the change, by offering to help them improve their own quality procedures.

In sum, the critical underlying question is whether breaking up the uniform, state-administered system of quality control will sacrifice standards. The fears are that inexperienced, cash-strapped exporters may cut corners on quality control costs, thereby sacrificing the high reputation of Ghana's cocoa for short-term profit – free-riding on the quality reputation of Ghana's cocoa but thereby undermining it. A counter-argument is that a licensing system to ensure that exporters are reputable, combined with statutory standards to be complied with, monitored by random checks, should safeguard quality standards. But the most reputable international trading firms are likely to be foreign. A common theme running through Ghana's agricultural marketing policies since Independence has been to replace the former foreign domination of Ghana's domestic and export trade. Therefore giving time to local firms to build their capacity and standards may be one reason why Ghana is pursuing a gradualist policy in privatising quality control and the cocoa export trade generally.

Ivory Coast: quality control is less successful[10]

Ivory Coast was one of the early liberalisers of its domestic and external trade. In Ivory Coast, unlike Ghana, there was never a state monopoly on cocoa trading, either domestically or externally. In the late 1990s it fully privatised the cocoa trade by abolishing CAISTAB, the state agency which had the official role of (i) regulating cocoa exports to avoid excessive sales at a given period that could adversely affect the market; (ii) guaranteeing effective delivery of the physical product; (iii) guaranteeing the quality of exports. Here we discuss briefly some of the problems with CAISTAB's quality control system. Quality control has now passed fully into the hands of the private exporters.

Under CAISTAB, control of quality passed through two broad stages. The first stage of control was carried out at farm-gates and warehouses across the country. At this stage, CAISTAB did not handle the product, the job being done by individual farmers or cooperatives. Quality control from the plantation to the warehouse (i.e. from fermentation to drying) therefore relied on the producers, or in some cases on itinerant traders operating on behalf of exporters. They inspected the product, authorised the sealing of sacks and the transport to nearby warehouses where the product should be fumigated.

The staff of CAISTAB intervened at the warehouses. They were trained and equipped to check the product and give a certificate to the buyer authorising him to transport the product to the ports. A second and final control operated at the port, where the exporter was obliged to grade the product by quality, with verification by CAISTAB staff. If the final decision of CAISTAB did not satisfy the exporter the process of inspection could be undertaken by private firms. With the abolition of CAISTAB, quality asssurance is now undertaken by the exporters or by private firms like SGS on their behalf.

To what extent does this new system affect the quality of cocoa exports? Before its abolition, CAISTAB had been withdrawn from inspection at the warehouses, before shipment to the ports. CAISTAB had argued that removing its role in internal marketing had damaged quality, since authorised moisture (2–10 per cent) never exceeded 5 per cent under the old system, but ranged between 5 and 10 per cent after CAISTAB inspections at warehouses were ended.

But it is doubtful whether the demise of CAISTAB affected quality of Ivorian cocoa, since the problems of poor quality control were long established – recognised by the market in the premium paid for Ghanaian cocoa over Ivorian. In Ivory Coast traders have tended to pay the same price for all beans regardless of quality. Beans have not been graded at farm-gates. Therefore cocoa growers have had no incentive to concern themselves with cocoa bean quality as is the case in Ghana. One major consequence is that beans of different quality have often been mixed. Exporters and traders could combat low quality awareness by rewarding growers producing high quality, e.g. with low interest loans, better sacks etc. But exporters have often been more concerned about shipment volumes. The big exporters like SIFCA or JAG now put emphasis on checking at the buying centres and commission private firms like SGS to undertake final inspection. Sub-standard beans are taken out of the export process. But smaller exporters have relied on the CAISTAB checkpoint at the ports. The challenge of raising Côte d'Ivoire's cocoa export standard is now in the hands of the private traders and exporters and their associations.

Zimbabwe: challenges and opportunities from market liberalisation

Zimbabwe has regulated extensively against animal and plant-related disease, both through phytosanitary controls on external trade and through checks on animals (particularly bovine diseases), seeds, milk and meat (when slaughtered and in butchers' shops). Enforcement is carried out by government officers, from the Department of Veterinary Services, the Department of Research and Specialist Services (DR&SS), the Department of Health, municipal governments and Customs (for border controls).

Meat inspection and seeds control were selected as case studies of how structural adjustment and liberalisation of markets have affected existing arrangements for quality assurance, and of the opportunities for alternative arrangements they have opened up.

Meat inspection and grading[11]

An inspection and grading system should be appropriate to the market served; it should also be low cost, reliable, comprehensive, and enable easy arbitration in disputes between animal sellers, abattoirs and meat traders. However, the service provided by government has come under increasing strain during structural adjustment. Liberalisation of the domestic meat market encouraged expansion in the number of small abattoirs (both urban and rural), while expansion of inspection and grading personnel has been restricted by declining real budgets in Veterinary Services and DR and SS. The result is that disease control by meat inspection in Zimbabwe's abattoirs and butcheries is thought by many to have declined in standard in recent years.[12] Inspection personnel in smaller abattoirs are regarded as having too little independence from the abattoir management. In addition, the functions of meat inspection and meat grading are carried out separately, requiring more personnel than if they were carried out by the same person, and there is some overlap of functions between health and veterinary departments.

Inspection personnel tend to be concentrated in the Cold Storage Company abattoirs slaughtering for the EU markets with its stringent disease regulations. Liberalisation of external trade in meat is expected soon, and is likely to stretch disease control services in abattoirs still further. High standards of meat hygiene are vital to meat export efforts, particularly in world markets; but lower standards may be appropriate to local markets.

Given the expansion in the meat processing industry, there is a clear need for change from prior arrangements in which inspection and grading were financed exclusively by the public budget. There are a number of alternative solutions to the meat inspection and grading problem:

- Increased budget allocations to enable recovery to previous standards of direct provision. But this is felt to be unrealistic given the constraints on public finances.
- Increased cost recovery, by making meat processors pay for the services provided by government inspectors, is the solution favoured by Veterinary Services and DR and SS for inspection and grading services – provided that revenues can be retained by the departments.[13]
- Merging meat grading and inspection functions – as in South Africa – is another way to reduce costs.
- Extending and formalising the practice in private abattoirs, which employ their own inspectors and graders subject to occasional checks from government inspectors.
- A more radical approach would be to contract out the entire inspection and grading function to a private firm, but this would still require finance and supervision of its operation.
- The option of setting up a stakeholder association (as in the former Meat Board in South Africa) to regulate and oversee the inspection and grading function would face the challenge of integrating various producer, processor and trade interests. It also runs the risk of control by more powerful players, with possible use of standards as a means of excluding other producers. A leading government role might still be required, even if there is a degree of partnership with the industry – especially in a meat exporting country, whose overseas customers (e.g. the EU) require government guarantees of quality.

The seed industry[14]

In Zimbabwe the seed industry has undergone rapid change and development under the Economic and Structural Adjustment Programme.

Government organisations, by both design and default, are transferring responsibilities for managing plant breeding, extension, multiplication, certification, marketing and quality control to private organisations. This is a common trend in many countries around the world. Research-based seed companies now dominate the supply of hybrid seed while International Agricultural Research Centres, NGOs, informal farmer groups and commodity traders are dominating the supply of non-hybrid seed.

But the government continues to dominate the supply of seed to farmers in marginal areas through drought relief programmes. Also, government organizations still carry out ad hoc monitoring of private companies particularly in seed certification, laboratory testing and issuing of International Seed Trade Association 'International Orange Seed Lot Certificates' and phytosanitary certificates for seed entering international trade. The problems in these arrangements include excessive bureaucracy and delay in getting seed analysis certificates from government departments; numerous changes to the Seeds Act that have not been gazetted; lack of consistency among government departments; and compulsory seed certification. These problems are constraining further growth and improved performance of seed markets. There is a need to redefine roles and put in place new relationships among public and private organisations to successfully manage the transition from an administrative to a market-based system.

Options and recommendations for future policy to manage the five critical areas in the public–private interface are as follows:

- *Germplasm maintenance and conservation*: these have both public and private good characteristics such as indivisibility, non-rivalness in consumption, jointness in production, high exclusion costs, uncertainty and non-appropriability. There is a public role that can be carried out by a specialised government department. However, expansion of government funding to provide adequately in this area is unrealistic, so the only feasible option is for seed services to leave germplasm maintenance and conservation to commercial companies, universities, international agricultural research centres (IARCs) and NGOs.
- *Plant breeding and varietal release*: the issue here is the future of the Crop Breeding Institute. The options are to continue with breeding all major crops, but with substantial increase in funding, to privatise the Institute completely, or to retain ownership of the Institute and concentrate its efforts exclusively on marginal areas and crops that

are not getting the attention of private breeders, particularly open pollinated grains. The last option is recommended.

- *Field and laboratory inspections for seed certification*: these services by government have deteriorated owing to budget cuts. Furthermore, the market has gone beyond current legislation and become sophisticated enough that consumers exercise control by deciding among different products and outlets. The only feasible option is to authorise an independent seed inspectorate financed by the industry to take over the registration of new cultivars produced and marketed in Zimbabwe, seed certification, monitoring of plant breeding research, and registration of seed inspectors and private laboratories, thereby taking this burden away from seed services.
- *Phytosanitary controls and issuing of seed analysis certificates for seed entering international trade*: the quality and timeliness of the Plant Protection Research Institute's services deteriorated drastically during the 1990s because of the depletion of skilled and experienced staff, lack of transport, poor equipment and supplies, inadequate funding, and poor administration and management. Given that privatisation is not possible owing to international regulations, the sensible option is for DR and SS to carry out phytosanitary services with full cost recovery.
- *Maintenance of seed security stocks*: given the lack of government funding, this should be left to private companies.
- *Training of breeders and technicians*: similarly, this should be left to private companies.

Milk testing

Where producers, processors and consumers require objective assurance of the safety and quality of a product, an association representing all of them (with government overview) can be an effective and lower-cost way to provide that assurance, provided business management in the association is strong.

Dairy Services, a department within Zimbabwe's Ministry of Lands and Agriculture, provides safety and quality tests on milk. With the deregulation of the milk market in the early 1990s, the number of milk processors increased from two to eight and Dairy Services, facing budgetary cuts, was so hard pressed that its continued operation was threatened. In 1993, the Zimbabwe Dairy Herd Improvement Association (ZDHIA) was set up by the milk industry with the approval of the Ministry of Agriculture to manage the milk recording service and aim for financial self-sustainability. The Dairy Industry Trust Fund was set up to manage a levy designed to reduce the costs of statutory milk recording. The ZDHIA 'joint venture' with Dairy Services thus enabled resources to be channelled from the industry directly to support the statutory service provided.

As a result of this process, Dairy Services moved from 97 per cent dependence on government funds for its operational costs in 1992–3 to 38 per cent in 1996–7. It was anticipated that government's contributions should amount to approximately one-third of total revenue, being the costs of subsidies on testing services to smallholders.

The Ministry of Lands and Agriculture had planned in the mid-1990s that Dairy Services would no longer be a government department. The thirty-four Dairy Service staff would be transferred to the newly formed Dairy Services Association of Zimbabwe (ZDSA), which replaces the former ZDHIA. The assets of the Dairy Service, mostly old but serviceable, would also be transferred to ZDSA. Private sector accounting and auditing procedures would be introduced. Government would retain ownership of the premises but ZDSA would pay for all maintenance and improvements. Government would retain a role in management of ZDSA through its membership of the governing council. Since the Dairy Act provides for the regulatory duties to be performed by designated persons rather than only by members of the public service, statutory requirements did not stand in the way of this transition to regulation by an industry association.

Advantages of the arrangement would include:

- no disruption of testing services caused by the transition
- government retains a major 'steering' role, while its resources are freed for other uses
- testing services will be market sensitive
- all jobs in the existing Dairy Services are retained on a full salary, pensioned basis while being removed from the government budget
- assurance of participation of all stakeholders
- assistance to smallholder dairy farmers can be provided.

The tight-knit nature of the dairy industry, in which grading and disease control play a crucial role, help this planned transition. Well-targeted aid played a key role too. A Canadian CIDA programme provided for key Dairy Services staff to be trained in the running of market-oriented milk testing services in Canada, helping ZDHIA to build up its business and management at a critical time.

Although the plans for taking Dairy Services out of the Civil Service accorded with the policy of Zimbabwe's Public Service Commission, the move was opposed at senior level within the Ministry of Lands and Agriculture and had not yet been achieved at the time of writing (2002).

Conclusion

Switching policy from curtailing to developing private sector involvement creates the need for public agencies to adjust their function and how it is carried out. The case studies demonstrate different aspects of the need for role adjustment by public agencies. They are summarised in Table 11.1.

Key issues raised by the case studies are:

What public function in quality control is there in liberalised markets? A study of quality control in coffee exporting (Gilbert, Varangis and ter Wengel, 1999) concluded that providing information to producers is the sole quality control function in which state or industry organisations can usefully supplement the private sector. But where health risks result from poor quality control there is a public quality control function which can be carried out by state and industry. The case studies of meat and dairy inspection in Zimbabwe illustrate this point.

Is there a need for continuing state involvement in quality control for export trade? Differences in standards for exports and the domestic market have long been a feature of the food trade in developing countries, particularly in phytosanitary standards applied to meat and fish. Recent US and EU legislation lowering permissible residues of pesticides in imports further raises the difference. The existence of higher standards in export markets, plus the dominance of overseas trade by large trading firms enforcing the standards, suggests that public involvement in quality control for exports may not be necessary. But there is an argument in favour of licensing to ensure that firms cutting costs on quality control for short-term profit do not enter the market and lower the value of the 'collective brand'. This may apply particularly where state policy, as in Ghana's cocoa export trade, is to use liberalisation as an opportunity for less-experienced local traders to enter the export market.

How can adaptation in state quality control organisations be achieved? The case studies of state intervention in quality control for the domestic market (India, Zimbabwe) suggest the need for adaptation to a stronger private and industry role. The original objectives of the Department of Marketing and Inspection (DMI) (to protect the Indian producer consumer from exploitation) may now be better served by removing restrictions on domestic trade which frustrate market development (see

Table 11.1 Overview of case studies of organisational arrangements for quality assurance

Country	Organisations	Function	Performance	Adaptation to change
India	Directorate of Marketing and Inspection (DMI), Bureau of Indian Standards (BIS)	Grading of agricultural for Produce domestic market	Mediocre	Quality assurance for exports now left to exporters
Ghana	Quality Control Department of Cocobod	Ensuring cocoa beans are to the required high standard from farm to ship	High standard but high cost	Preparing for privatisation and competition
Côte d'Ivoire	CAISTAB	Quality control of cocoa at port, formerly at collection depots	Poor	Abolished in late 1990s
Zimbabwe	Department of Veterinary Services, Department of Research and Specialist Services (DR and SS), City Councils	Meat inspection and grading	Historically strong but deteriorated in the 1990s	Poor adaptation to market development
Zimbabwe	Seed Company of Zimbabwe, DR and SS	Seed, certification, germplasm conservation, plant breeding	Historically strong but deteriorated in the 1990s.	Poor adaptation tion to market development
Zimbabwe	Dairy Services, DR and SS	Testing of milk	Strong performance	Ready to move to full commercialisation but delayed by opposition within DR and SS

Chapter 3). Fraudulent use of its AGMARK label, overlaps with other public quality standards (BIS, ISI), and poor coverage of rural areas, suggest the need for review of the public role in quality assurance. The

DMI's energies should perhaps be focused on extending quality control awareness and standards more widely. In Zimbabwe, slow adaptation by public providers of quality control services to the needs of a growing market resulted in abattoirs making their own inspection and grading arrangements, and in failure to facilitate industry regulatory bodies in milk testing and seeds certification – despite the public service being run down in the latter case.

12
What Public Role is There in Market Information?

Recent market development efforts in developing and transition economies place much emphasis on the state's role in improving the environment for business. Provision of market information is often viewed as a priority, because adequate information about buyers, sellers and prices is usually lacking in poorly developed markets. In some countries there are long-standing market information services (MISs) run by government. Others have been set up recently with help from donors. In the light of theory and current developments, this chapter reviews briefly market information systems run by government departments in Ghana, India and Sri Lanka, plus one initiative by a farmers' representative organisation in Zimbabwe.

The main observations are, firstly, that opportunities for rapid increase in market information have emerged from the internet, mobile phones, market development and particularly from liberalisation of local broadcasting; and secondly that the main contribution governments can make to better market information is facilitating the spread of communications infrastructure and private local broadcasting.

Theory and current developments

The case for public provision of information is that it is non-subtractable and that there are substantial external benefits from accurate information in the form of less risk and better-informed decisions by farmers, traders and consumers. Consumption of non-subtractable (non-rival) goods does not reduce their availability for others. It is in the interests of each to understate their wants for such goods and to free-ride on the purchases by others. Where production of the good is not costless, left to itself the market will produce less than is required.

But in the case of information, the internet and radio have reduced the marginal cost of dissemination towards zero, vastly increasing the supply. Information (e.g. weather forecasts, stock market prices) has become a free by-product of advertisements for organisations and goods on the internet and local radio. The overall effect of developments in the internet, radio and telecommunications is to reduce market failure in information.

Market information services

A market information service (MIS) is designed to improve market transparency for all actual and potential participants in crop markets. By providing regular, detailed information on producer, wholesale and retail prices for specific locations, a market information service reduces the opportunities for non-competitive behaviour. Full information places all participants in a position in which they can negotiate prices. Furthermore, by making traders aware of the opportunities for profitable trade in other locations long-distance trade is promoted thereby increasing spatial integration of food markets and evening out food availability in surplus and deficit areas. Finally, by encouraging intra- and inter-year storage, temporal variability in supplies and prices is reduced.

If a market information service is to be effective, data must be available on a continuous and reliable basis and must be accurate. Prices must be collected frequently, preferably on a daily basis, specified for time and location, and processed and disseminated quickly via an appropriate communications network.

As part of their general statistics governments routinely collect data on agricultural products traded – at least the overall volumes. In countries where government gives priority to stabilising grain prices for food security purposes (e.g. India), more detailed data are collected and analysed, such as price variation between major population centres and seasonal price trends. In the 1970s and 1980s regional food security data collection (e.g. Horn of Africa, southern Africa) increased as donors and governments emphasised famine prevention as a result of major droughts. Agencies were created (e.g. the SADC food security unit) to assemble and analyse the data. This data is produced for planners and decision-makers primarily.

Where there is abandonment of crop price fixing by governments there is increased demand from the agricultural sector for reliable price information and price forecasts. Commodity exchanges and major traders in commodities advertise their prices in order to attract business, and the 'price search' for established commercial farmers is usually facilitated by market intermediaries. It is for the smaller farmers in remoter areas that market information is often thought to be lacking, particularly in newly liberalised market environments.

With the beginnings of liberalisation of agricultural trade in developing and transition economies, public initiatives began to produce market information for farmers and traders. The UN's FAO in the 1980s and early 1990s helped to set up MISs as a public service in several countries (e.g. Zambia). The UNDP helped develop Sri Lanka's market information network.

The case studies discuss public initiatives to provide agricultural market information to the private trade in Sri Lanka, India, Ghana and Zimbabwe. The case studies cover widely differing market information initiatives. In Sri Lanka and India the market information service was developed as an extension of agricultural marketing statistics collection for planning purposes. The Zimbabwe study is of a pilot market information initiative by an association of small farmers (ZFU).

Ghana: limited effectiveness of government's market information[1]

In Ghana the market information service was set up in the Ministry of Food and Agriculture (MOFA). The MOFA provides freely accessible information on market prices through the mass media to the market operators on a weekly basis. An evaluation of the service conducted by MOFA indicated that a number of operators considered it a useful guide on food crop prices. We also found that a number of urban traders were aware of the service but tended to depend more on informal market price information sources. The information is assembled by field staff employed by the Ministry. Limitations stem from communication difficulties which have delayed information flow between field officers and regional and national coordinating centres. The weekly publication schedules also meant that the information could prove out of date by the time of publication since prices tend to be quite volatile. This might explain why market operators tended to depend more on informal information sources than on the MOFA information service.

Data supplied on crop output also serve as early warning indicators on the supply situation for food security considerations. At the time of the study in September 1995 the MOFA had predicted a major glut in the domestic maize market. Government responded by encouraging banks to raise finance to buy off the excess crop. Bureaucratic delays within the banking system meant that by the time the decision was made the market had adjusted to the crop supply situation and the anticipated glut did not materialise (ISSER, 1996). This anecdote puts a question mark over both the need for a public price information system, and the capacity of MOFA to operate one effectively. Quite a large data base on food crop output and prices had been built up but it appeared that this was not being subjected to much analysis to guide policy.

India: a market intelligence scheme by government mainly for government[2]

The market intelligence scheme operates as one of the services under the Directorate of Economics and Statistics (DES), part of the federal Ministry of Agriculture. The objectives of the DES are improvement in quality, coverage and timeliness of agricultural statistics required for decision-making on various policy issues such as procurement, prices and import/export of agricultural commodities. The DES undertakes the following activities:

(i) collection of statistics on area, production, yield of principal crops in India; cropping patterns; agricultural wages; as well as collection of wholesale, retail and farm-gate prices of agricultural commodities;

(ii) processing of this data for the purpose of policy framing, publication and monitoring of prices;

(iii) dissemination of data mainly through printed media and in some cases (as with daily prices), through radio;[3]

(iv) undertaking research and analytical studies on the basis of information collected by the Directorate as well as preparing policy documents and background papers for internal use and for various committees established by the government from time to time.

The market information system was introduced in the mid-1950s as one of the activities of the DES. In recent years DES has concentrated on the assembly, processing and dissemination of prices data (weekly, bi-weekly and monthly) as well as other agricultural statistics. Each state is responsible for the collection of data from important wholesale and retail markets within their jurisdiction, the reporting of this to the Directorate, and the dissemination of data within its own region. The staff of the State Agriculture Departments located in the field are specifically charged with the task of reporting the daily prices directly to the DES.

The DES also collects data directly through its established network of Regional Market Intelligence Units in fourteen states.

The DES's objectives in data-processing are wider than that of the states, since their data forms the basis for formulation of food and agricultural pricing, procurement, export–import and credit policies by the concerned ministries/agencies/Planning Commission. Agricultural prices are monitored at the highest level in the government. Fortnightly meetings on agricultural prices are held in the Cabinet Secretariat with representatives from the Food, Agriculture, and Civil Supplies, Consumer Affairs and Public Distribution Ministries to review food availability and plan for any emergency.

The dissemination of data is carried out at two levels, centre and state. The main medium of communication by both bodies is the state-owned All India Radio. The DES confines itself to national broadcasts of daily wholesale and retail prices of important agricultural commodities such as rice, wheat, coarse cereals, for fifteen to twenty important wholesale and retail markets, those located in high consumption zones with very high purchasing power. At the state level price data is broadcast daily, in regional languages and in rural programmes.

If a market intelligence scheme is to be effective data must be available on a continuous and reliable basis and must be accurate, specified in terms of time and location, and processed and disseminated quickly via an appropriate communications network. These criteria immediately highlight the strengths and weaknesses of the market intelligence scheme managed by the DES.

A strength is the coverage of the vast agricultural market in India, with its heterogeneity, geographical dispersion and segmentation. Data on wholesale and retail agricultural prices is collected from more than 3000 markets all over the country by different states. A further strength is its adaptability. Over the four decades of its operation it has increased considerably in scope, changed its structure and decentralised its operations. The delegation of data collection responsibility to the states enabled expansion in its geographical coverage and its data-base.

Weaknesses are the gaps in the dissemination process. In the mid-1990s there were several regions, like the north-eastern region, and some tribal and hill regions, which were not adequately covered by the scheme. Broadcasts can be sporadic and subject to frequent disruptions mainly because timely supply of information from the state departments was not always ensured. The weekly *Bulletin on Agricultural Prices* was produced with a significant delay of between three and four months. While the Directorate produced daily price information this was for internal use only. Prices were not broadcast on the television network by the DES.

A crucial issue is whether the government's MIS duplicates private information. There are numerous privately owned newspapers which provide daily wholesale and retail prices of major agricultural commodities. The source of information for the newspapers are the market committees of the wholesale/retail markets. This suggests that the users of DES market information are in the rural areas where radio still remains the exclusive medium of communication, and that the public and private systems complement each other.

Sri Lanka: a research institute based market information system[4]

Among the organisations involved in the market information system of Sri Lanka the key role is played by the Marketing and Food Policy Division (MFPD) of the Agrarian Research and Training Institute. Its efforts are supported by the Prices and Wages Division of the Department of Census and Statistics, the Department of Agriculture, the Provincial Directors of Agriculture, and the Sri Lankan Broadcasting Company.

The MFPD processes data into a weekly Food Commodities Bulletin and a monthly Food Information Bulletin, and disseminates data via radio, television and newspapers.

A market intelligence system has been operational in the Agrarian Research Training Institute since 1979, assembling and disseminating information on production, consumption and prices of food products on a weekly, and in some areas daily, basis. Initially, the MFPD collected retail and wholesale prices in the Colombo area and ten other locations in the country for a selection of food items. By the mid-1990s – after expansion with UNDP assistance – thirty locations were covered with the goal of eventually including a total of seventy five locations. Commodity coverage is wide with many sub-commodity categories.

The use of local radio stations for dissemination allows the system to reach the farming community which has less access to printed media. It also permits the data to be targeted to relevant users and reduces the amount of data having to be transmitted to a central location at the same time.

On the supply side, the market information system was working well. Lacking at the time of the research was solid evidence on how the data disseminated by the MFPD is received and utilised by the private sector. But the fact that a private fertiliser company enquired about the possibility of transmitting its name immediately after the price broadcasts suggests that it believed its target audience could be reached in this way.

In summary, the MFPD has demonstrated its capacity to maintain a price collection service, while the proposal to UNDP to help expand and improve the existing service came from within MFPD itself, indicating the commitment of the staff and a recognition of the weaknesses in the pre-existing network.

Zimbabwe: a market information trial by a farmers association[5]

The Zimbabwe Farmers Union represents small farmers, mainly from the communal areas. The Zimbabwe Farmers Union's pilot MIS in the mid-1990s illustrates the challenges facing the establishment of a successful public MIS by an association for small farmers in remote areas, who have multiple marketing constraints.

ZFU's pilot MIS was a response to the changing information needs of farmers in a liberalising economy. Prior to liberalisation, farmers had little use for market information since prices were fixed and government bought all produce offered to depots. The signals on what and how much to produce were provided by government pricing policy.

Liberalisation of the economy has resulted in changes which affect the need for and use of market information by smallholder farmers:

- removal of restrictions on the movement of previously controlled crops (mainly maize) potentially gave farmers the opportunity to sell further afield;
- closure of the less financially viable state depots increase distances from farm to depot, making grain sales to state depots less attractive.

But liberalisation has not resulted in a surge of private trading in Zimbabwe's communal areas. Remoteness, the small quantities sold by individual farmers and the semi-subsistence orientation of poor communal farmers causing them to sell only when they have need for cash – all mean that itinerant buyers are the main traders. A result of this type of marketing is that discussions of purchase prices at any one time are undertaken between a single seller and a single buyer in those cases where sellers approach buyers; or a single buyer and many sellers in those cases where a buyer goes into an area. Lower prices result because there is:

- absence of direct competition among buyers as buyers are rarely in an area simultaneously;
- uncertainty among farmers about when the next buyer will be in the area and what prices that buyer would offer.

The objective of the ZFU's trial MIS was to help the smallholder farmer:

- to be adequately informed and capable of keeping abreast of reforms;
- have timely access to required information in order to make timely decisions on what to grow, where to sell and when to sell it (ZFU 1994).

The major concern was to improve the bargaining position of farmers by access to information. This required the MIS to concentrate on the transmission of information about different sales outlets and the ruling prices in these outlets for both input and output markets, plus estimates of transport charges involved in accessing the different outlets.

The MIS pilot was financed by ZFU with the help of a grant from the Swedish Cooperative Centre and operated through existing ZFU structures. The ZFU is represented at the national, provincial, district, area/ward and village/group levels. At the national level, information was printed on leaflets and made available to the local branches. In total, about 50 000 leaflets were printed monthly, with information on the markets for grains, oilseeds, cotton, livestock, occasionally horticultural produce, fertilisers, seeds and grain bags. The focus was on prices in specific markets, usually state marketing depots where the price is similar throughout the country or outlets in Harare, or prices on the Zimbabwe Agricultural Commodities Exchange (ZIMACE). In addition to the English language leaflets, the information was processed for a national radio broadcast in the vernacular languages Shona and Ndebele.

The study assessed early consumer reaction to ZFU's pilot MIS, by interviewing farmers and traders in a number of locations, in different natural regions and with different accessibility to major centres. Overall the study found that little direct use was being made of the information distributed. Interest and communication among farmers was predominantly about local market issues, such as acceptable moisture levels in grain, delays in payment, transport availability and costs and grading reliability. Farmers' main local information sources were other local farmers, extension agents, local leaders, and well-travelled family members. Information about prices in what are seen to be remote markets beyond their reach was felt to be of less interest by the farmers. Farmers interviewed felt that the market information obtained from existing radio broadcasts was often general and focused on distant markets where participation would involve long travel and payments for accommodation.

In sum, the overall finding was that producing and improving specifically local market information and keeping it up to date is not a task to which a national MIS run centrally can easily be adapted. Local radio is the most promising medium, but is not yet established widely in Zimbabwe, and broadcasts would need to be locally sourced. This would pose a challenge for a small farmers' organisation to finance – it cannot be financed by subscription – and manage on a decentralised basis.[6] Furthermore, any information system for small farmers in Zimbabwe should arguably include more than market prices, since the binding constraint on market integration for small farmers appears not to be information but finance, organisation, transport and the assembly costs of multiple small lots on scattered holdings.

Conclusion: what state role in market information?

The case studies above, and experience elsewhere with MISs (e.g. Zambia), suggest MISs function better when they have become part of a well-established agricultural statistics service, as in India and Sri Lanka. Where statistical services were run down (e.g. Ghana) or where the MIS was started by an association (e.g. ZFU in Zimbabwe) the challenges of sustaining a good service are great. Even so, government information systems tend to serve government needs, particularly food security policy, rather than being tuned to the needs of farmers and traders:

- Information is often too general and out of date. Buyers and sellers need today's prices for their area;
- When farmers face multiple marketing obstacles (e.g. transport, cash to pay marketing expenses) information – even when potentially useful – cannot necessarily be acted upon;
- In well-developed agricultural markets, private provision of current market information predominates (disseminated by suppliers, marketing associations and exchanges) with public information focusing on price and quantity trends countrywide, or regionally;

- None of the systems monitored the usefulness of the MIS to farmers and traders. Information regarding usefulness is patchy, but indicates that they are not usually the main source of information for buying and selling decisions;
- Where government needs to take on the role of collecting, analysing and disseminating current market information to assist the private sector, the underlying problem may be that private trade, exchanges and broadcast media have been hindered in their development (e.g. by government monopolies). In these circumstances a key contribution which government can make to better market information is to remove regulatory and bureaucratic obstacles to private provision of information.

These observations accord with an assessment by FAO,[7] which points to opportunities offered by local FM radio stations. These are being taken up in Mali and Uganda, though more slowly in Ghana and South Africa despite the existence of local FM stations. Organisation and finance remain a problem. Existing services usually have donor or NGO finance (as in Mozambique), and the radio stations treat the market information broadcasts as paid business. To make the services sustainable they need to find commercial sponsorship, which should not be difficult once the stations attract the attention of advertisers.

In sum, market information services for farmers, traders and consumers rely increasingly on radio stations and the internet, and are financed by susbscriptions (for specialised associations), NGOs and donors, and by commercial advertisers. The role of government in market information services seems likely to diminish, with state services focusing on collection and analysis of data for policy purposes rather than trying to fill a dual role as in the past.

Part IV
Conclusion

13
Developing Agricultural Trade: New Roles for Government

State trading and restriction of private agricultural trade became virtually the norm in many countries from the mid-twentieth century, particularly in the grain trade, owing to distrust of private trade, and policies of national self-sufficiency in grain and protection of domestic agriculture from foreign competition. High costs of state trading and public stocking of grain, plus the perceived benefits of market development, have swung policy preferences towards reducing trade restrictions and developing the private trade. Better trade information, quality assurance through branding and traceability, and the emergence of futures and options markets for the trading of risk are features of market development in agricultural trade. Market failure in the sector is thereby reduced. This provides the opportunity for government to shift its role in agricultural trade from buyer and seller to provider of infrastructure, facilitator of trade development[1] and regulator of trade in the public interest (providing a legal framework, ensuring competition, consumer safety and protecting natural resources).

However, market failure has rarely been the only reason for government intervention in agricultural trade: political interests come to the fore in any change in trade intervention and affect government decisions. Beneficiaries of protection and state trading – farmers, monopolistic millers, state trading employees – resist removal of protection. International trading companies argue for it. Furthermore, as Bates (1989) noted, deregulated markets adjust rapidly, leaving reorganisation of government institutions as the real task of structural adjustment. Hicks (1969) went further, questioning whether the constant unmaking and remaking of production structures intrinsic to capitalism could ever be compatible with the traditional societies dominant in developing countries. Yet social, political and market development are all now on the agenda in developing countries as they seek to integrate into global trade and investment. The challenge for governments is to adjust their role to enable their people to reap the benefits of market development.

This book has explored the varying approaches, conflicts, capability and will of governments with regard to opening their agricultural trade and adjusting the public services they provide to agricultural trade. In this final chapter the argument developed is first summarised.[2] Reforms in agricultural trade and related public services in the case study countries are then overviewed, and a conclusion drawn.

Summary of the argument of this book

First, after many decades of policies emphasising state control of agricultural trade, developing the private sector role emerged from the 1980s as the dominant goal of public policy in most countries.

Second, there is an increasingly active private sector in agricultural trade. This is partly[3] because governments are liberalising, to varying degrees, their external and domestic trade in agricultural products and inputs. Trade in staple grains has tended to be liberalised least and last. But liberalisation of foreign exchange markets – allowing private individuals and companies to hold and deal in foreign exchange – may have contributed more to increased private sector activity than has removal of restrictions on trade. Liberalisation of trade has been a policy goal and process rather than a once-and-for-all accomplishment. Removal of official restrictions on trade and reduction of state buying, selling and storage has proceeded unevenly in most of the countries studied, because of conflicting interests involved. It has often taken place only as a result of crises in public finances which tip the balance of decision-making towards trade reform, encouraged by donors. It is ongoing in some countries and sectors, while stalled or in reverse in others.[4] But it remains the policy goal in most countries, as a means of promoting a market economy. The rapidity with which countries signed up to WTO, which is committed to achieving freer trade, demonstrates this point.

Third, the rise in private trade has increased the mismatch between existing public services for agricultural trade (often designed for control by the state) and the public service needs of consumers, producers and traders of agricultural products and inputs. Where state companies and departments had formerly monopolised trading (e.g. abattoirs and maize in Zimbabwe and Kenya) the entry of private companies able to trade at lower cost left state facilities under-utilised. Where states operate expensive universal rice ration schemes (e.g. in India, formerly in Sri Lanka and Bangladesh) the private trade has often been able to sell rice more cheaply than the state.

Fourth, greater capacity to manage reform of public services is revealed in more consistent reform, sustained quality of public services, and increased collaboration with private firms in providing public services, and with donors where aid is needed. There is some evidence from the case studies that capacity to manage reform differs by country (greater in India and Sri Lanka, lower in Ghana, Kenya and Zimbabwe). This capacity to reform does not result from greater wealth. Bangladesh, one of the poorest countries, is widely cited for its success in managing its shift to targeted rice rations, achieved despite limited resources in the state, by using the collaboration of donors (Ahmed, Haggblade and Chowdhury 2000). But cutting across the classifications of reform performance by country are pronounced differences in reform performance between trade in grains and non-grains: reform of public organisations and their services seems least successful in uncompetitive staple grain industries (e.g. maize in Kenya, rice in Sri Lanka) and most successful in non-grain export industries (e.g. Cocobod in Ghana).

Where staple grain industries are uncompetitive, food security concerns and pressure from farmers for protection against imports often hinder liberalisation of external trade and reduction of state participation in buying, storing and selling grain. In Sri Lanka and Kenya farmers' protests over price falls associated with grain imports led to attempts to re-establish state price support. By contrast, liberalisation of the rice trade in Vietnam has helped it become the second largest rice exporter, without deprivation of the poor. Equitable distribution of land and knowledgeable small farmers are argued to be a key factor enabling small farmers to participate in the benefits of liberalisation (Minot and Goletti 2001).

Successful exports are by definition competitive. The infrastructure they create in order to move export crops from the hinterland is therefore economically robust. By contrast, the infrastructure of depots and transport set up by public grain agencies buying grain in remoter areas at subsidised prices has been the first casualty of commercialisation of these agencies, thereby disadvantaging grain farmers in remote areas (e.g. Kenya). This is a further reason why reform of the state role in export industries (e.g. cocoa in Ghana, dairy products in Zimbabwe) may be accomplished more easily.

Finally, competitive contracting (the heart of the 'new public management') has been little adopted as a means of reforming public services to agricultural trade, even when governments have adopted commercialisation and contracting as policy, and actively contract-in services for government (e.g. cleaning, catering), as in Zimbabwe. Where corporatisation[5] of public agencies has been adopted it has not proceeded to the point where a corporatised agency tenders competitively for public work.[6] Capacity to manage tendering and monitoring of contracts does not appear to be the reason, since using private contractors (e.g. for capital investments and supplies) is long established in governments. The finding of this study is rather that unwillingness to reform organisational arrangements is the reason, usually owing to lack of a sufficiently powerful political coalition for reform of public services, able to manage the short-term trade-offs among winners and losers from reform. Management capacity in public service commissions and ministries is important to the reform process, but it cannot carry through conflict-ridden reform without sustained political support.

As a result, where corporatisation has been adopted it is superficial (e.g. in the Grain Marketing Board in Zimbabwe, and National Cereals and Produce Board in Kenya), and where there is supposed to be competition for government work between independent government agencies and private providers there is in practice very little (e.g. State Warehouse Corporation in India). Programmes in Zimbabwe to contract out some tasks in government veterinary work and tsetse control were initiated with donor support in the late 1990s but were stalled by the subsequent breakdown in government–donor relations.

While NPM has not been adopted other than in name, there is increasing private provision of services formerly provided only by the state. Some services formerly provided by state agencies were not public goods (i.e. they could be financed and operated by a sufficiently developed private sector). So when private trade became stronger, helped by liberalisation, and where government services became weaker, the private trade took up provision of the services (e.g. meat hygiene inspection in Zimbabwean abattoirs). There has also been some privatisation (e.g. milk and cotton marketing in Zimbabwe, cocoa buying in Ghana).

Private finance and management have entered into public services through collaborative arrangements between government departments and private firms (e.g. the rice bondsmen and the PRIMA wheat milling contract in Sri Lanka, and the new government incentives in India for private provision of grain storage). Such collaborative arrangements are more common in countries with stronger capacity in government, perhaps owing to greater policy stability creating more confidence among investors.

Figure 13.1 sums up the above points, illustrating how the analysis leads to the main issue considered: how governments reform or fail to reform public services to agricultural trade.

CHANGE IN IDEOLOGY after many decades of policies emphasising state control of agricultural trade, dominant ideology changed towards developing rather than suppressing private trade

A STRONGER PRIVATE SECTOR IN AGRICULTURAL TRADE emerges, to varying degree, often more the result of liberalised foreign exchange markets than liberalisation in agricultural output and input markets

MISMATCH EMERGES BETWEEN EXISTING PUBLIC SERVICES AND NEW PUBLIC SERVICES REQUIRED BY AGRICULTURAL TRADE as state services which compete with private traders become redundant, and as the need increases for new public services to facilitate market development, particularly in less developed areas

GOVERNMENTS WITH LOWER CAPACITY[7]
Inconsistent liberalisation of agricultural trade and reform of public services. Public services decline. Some private and NGO provision of public services by default.

GOVERNMENTS WITH HIGHER CAPACITY
Liberalise agricultural trade and reform public services more slowly and consistently. Increased collaboration with private firms in providing public services

COMPETITIVE CONTRACTING IN PUBLIC SERVICES REFORMS IS NOT GENERALLY ADOPTED. But there is increasing private and NGO provision of services formerly only provided by state.

REFORM IS EASIER IN EXPORT INDUSTRIES AND MORE DIFFICULT IN UNCOMPETITIVE STAPLE GRAIN INDUSTRIES, therefore reform performance in an industry may be different from overall public services reform performance of a country (e.g. reform performance in the public role in Sri Lanka's rice trade has been poor despite the strong capacity of government in other sectors).

Figure 13.1   Sequence of argument

Table 13.1 Overview of agricultural trade liberalisation and reform of public services to agricultural trade in case study countries

Country	Liberalisation of agricultural trade [a]	Reform of public services to agricultural trade	Political coalition for reform of public agricultural services	Donor involvement	Capacity for managing public services		Collaboration with private sector in providing public services	
					Services policy development	Implementation of services	Use of private finance and management	Use of competitive contracting
Ghana	Domestic trade in cocoa and part of export trade in cocoa	Government trading and stocking of maize ended	Strong	Much	Weak	Weak (except in cocoa sector)	None	None
India	Very little in domestic or external grain trade	Minor reform in Public Distribution System	Weak – but now (2002) strengthening	Little	Strong	Strong	Beginning	None

Kenya	Domestic trade in grains; maize imports	Attempts to corporatise, but little change	Weak	Intermittent	Weak	Less weak	Little	Little
Sri Lanka	Import of rice	Government trade in rice ended; rice and wheat import contracts begun	Strong None	Much	Quite strong	Quite strong	Quite strong	None
Zimbabwe	Domestic trade in grains; wheat imports[b]; cotton exports	Successful privatisation in dairy and cotton; attempt to corporatise grains but little change. Services deteriorating	Weak, and disrupted by political upheaval	Intermittent	Less weak	Weak	Little	Little

[a] Foreign exchange markets have been liberalised for trading transactions in all the case study countries. This has substantially increased private trade in agricultural commodities and inputs in some countries.

[b] All private trade in grains in Zimbabwe is currently suspended (2002).

Reforms in the case study countries

Reforms in the case study countries, plus aspects of governments' ability to carry out reforms – strength of the political coalition for reform, donor involvement, and capacity to manage public services – are overviewed in Table 13.1. Also included are indicators of collaboration with private firms in providing services (use of private finance and management, competitive contracting). The summary comment in each cell is over-simple, but emphasises points of difference between countries.

Ghana

Reform has focused on the export sector, notably Cocobod, which formerly held a monopoly of cocoa, coffee and sheanuts trade externally and domestically. Liberalisation has been slow but substantial. Ghana's reputation for quality cocoa has been maintained while farmers are receiving a higher percentage of the export value. There is a strong coalition for reform, and close collaboration with donors. The civil service and policy process at all but the highest level of government are relatively weak. Rural infrastructure is weak, as is management of natural forests, which are reduced by expansion of cocoa production. In grain markets, government ended its minor role in buying, selling and storing maize, but has yet to sell off the storage facilities of its non-operational public grain marketing agency. National food security is not a major issue in public sector reform, owing to the diversity of diet, plus increases in maize and rice production, all privately marketed.

India

Until recently there has not been a strong coalition for reform.[8] Despite substantial liberalisation of trade in manufactures and services since structural adjustment began in 1991, there has been virtually no effective liberalisation of agricultural trade or reform of public organisations in agricultural trade. The civil service and policy process is relatively strong. Donors play only a minor role. Food security is a continuing national priority. Despite evidence of massive inefficiency in public grain distribution, governments have chosen to pay the costs rather than reform. There is now much innovative collaboration with the private sector in providing facilities at transport terminals. Such public–private collaboration appears to be the direction of reform (e.g. freeing of the Central Warehousing Corporation to invest abroad), rather than outright privatisation of agencies.

Kenya

Reform has been uneven, driven by successive crises in public finances, with much backtracking – particularly in reform of the public grain marketing agency. While there is more success in liberalising non-grain agencies (e.g. milk and tea) political interference remains a problem. A strong political coalition for reform has been lacking, and the working relationship with donors has been conflictive. Domestic grain marketing was liberalised in the later 1980s, followed by maize importing in 1993. Intended ending of public maize purchasing to support prices has not occurred owing to pressure from maize farmers, as maize imports have lowered prices. An import levy on maize has been raised to protect farmers. Excess maize is exported at a loss after bumper harvests of good rainfall years. Rural infrastructure has declined as public finances and services have weakened.

Sri Lanka

There has been limited liberalisation of external trade in grains and abolition of the former universal rice ration, followed by ending of state buying, selling and storage of rice. There have been collaborative arrangements with private importers of rice and flour millers to provide a national food security rice stock at zero governmental cost, and free milling of government imports of wheat. Civil service capacity is relatively strong in policy-making and implementation. Political coalition for liberalising reform was strong up to the mid-1990s. There has been continuing collaboration with donors on reform policy, since structural adjustment began after the financial crisis of 1979–80. Rice policy remains unsteady owing to the conflicting demands of rice farmers and cities. Traditional rice production in Sri Lanka is uncompetitive with imports, resulting in falling prices as imports have risen, and increasing pressure from farmers for public intervention. The stresses in food policy were revealed in the mid-1990s when subsidies to wheat imports (using food aid) to reduce consumer prices were combined with revived state buying of rice to maintain producer prices. A variable levy on rice imports has now been imposed.

Zimbabwe

Domestic trade in grains was liberalised in the 1980s, in addition to wheat imports and cotton exports. Dairy and cotton marketing agencies were privatised with introduction of limited competition. But attempts to corporatise grain marketing have not succeeded owing to continued political interference. The perception that control of maize markets and imports is essential for food security remained strong among policy-makers, even before the state-sponsored invasions of commercial farms. All private trade in grains is currently suspended. The political coalition for reform has been very weak, and the previously strong public services have deteriorated with successive financial crises and loss of their former managers. Donor involvement has been intermittent and ridden by conflict with government, as in Kenya. Government antipathy to a private sector dominated by white-owned businesses has slowed reform. Long-term overcrowding and lack of development in communal rural areas has fuelled political tension. Zimbabwe faces a major challenge in developing rural livelihoods.

Conclusion: the development challenge underlying public services reform

The 'new public management' (NPM) has focused the debate regarding reform of public services on to how services are carried out rather than what public services are provided and for whom. The main finding of the study concerning the relevance of the NPM model is that it has hardly been taken up as a model for reform of services to agricultural trade, despite policy rhetoric in some countries. NPM-style reforms with their stress on competitive contracting may become more feasible and attractive as markets develop.

But when applied to public services to agricultural trade, as in this study, the question of whether public services are provided internally within government or on contracted out basis emerges as relatively minor. The priority is to shift public resources out of public organisations and activities which hinder market development and into those which enable wider beneficial participation by farmers, traders and consumers in markets which are expanding and increasingly open.

The development of private trade reduces market failure in the sector, thereby moving the boundary between public and private goods, and enabling government to shift its resources as required. But many governments lack the capacity to make the changes, particularly the political authority. The greatest challenges in accomplishing this shift are where public marketing organisations have supported uncompetitive production and market development is least. The potential conflicts involved with liberalisation of trade in these cases (particularly where staple grain production is uncompetitive) has been a key reason for lack of political resolve to reform. In many cases competitiveness is largely a dynamic matter, given that so much of the value added is downstream from production. Therefore reducing the downstream costs in transport and handling (e.g. by providing better infrastructure), enabling finance to flow more easily into trade (e.g. by encouraging use of warehouse receipts as collateral), and helping to develop the markets required for more efficient trading (e.g. by encouraging futures and options trading) can raise competitiveness.

Market development in remote areas presents a greater challenge in leadership and collaboration of government, private and voluntary sectors. Significantly, the governments which find reform of public intervention in agricultural trade the most difficult to achieve tend to be the ones least involved with the task of developing rural areas. Shifting the focus of policy effort and debate to rural services may pay long-term dividends in easier reform of the state role in agricultural trade.

Notes

Preface

The collaboration involved five British research groups: the International Development Department, University of Birmingham; the Overseas Development Group, University of East Anglia; the Health Policy Unit, London School of Hygiene and Tropical Medicine; the Water Engineering and Development Centre of Loughborough University of Technology; and the Department of City and Regional Planning, University of Wales. Richard Batley of the International Development Department, University of Birmingham provided coordination.

1. Governments and Markets

1 Economic theory advocates that efforts to redistribute income should not interfere with markets for goods and services, so that redistribution does not alter relative prices and thereby affect private allocation decisions. Redistribution should be through giving assets – ideally money, in theory, since it gives the beneficiaries the widest choice of use – in a manner which reinforces productive use of those assets, e.g. 'workfare' to able people in need, temporary cash support for an infant industry, rather than tariff protection.

2 A three-part programme for introducing a greater private sector role into animal health services, recommended by Umali, Feder and De Haan (1994), provides an example of application of the market failure approach. First, services which are basically private should be moved to the private sector. To facilitate this move, both removal of barriers to private sector involvement and legal underpinning for private agreements are needed. Second, for services 'that entail externalities, moral hazard, or free-rider problems, mechanisms to correct these market failures are needed' (ibid.: 93), otherwise the private sector will either underprovide or overprovide such services. Third, where services are regarded as essential for the poor, but will not be provided by the private sector, subsidised, targeted public provision is recommended.

3 Put in formal economic theory terms, well-functioning markets will tend to use more or less of a resource according to whether its marginal value product is greater or less than its marginal cost – provided the 'rules of the game' allow them to do so. In so doing they generate income. If the 'rules of the game' block such adjustment in resource use prices will not represent market prices (i.e. equilibrium prices), at the cost of income generation. When 'rules' are far out of line with market opportunities, relative prices of resources will tend to be far from their market levels, and those suffering the cost may be aroused to collective action to overthrow the rules (North 1990). Thus is social change brought within the economic calculus by new institutional economics.

4   Associated with the early twentieth-century work of Commons and Veblen.
5   This leads to the question whether a particular form of institutions is most market-friendly (e.g. property-owning democracy). If so, that form will be the most responsive to changes in market opportunity, so that markets and institutions will be kept in step by incremental change rather than upheavals. Such institutions may be needed for an open economy to work, since markets change more rapidly in a liberalised economic environment. The state has a role in balancing and smoothing the process of institutional change – to do that it must be devoted to the interests of the majority, to well-functioning markets, and willing to engage in a process of continuous institutional development.
6   Walsh (1995: 24) suggests they are not inherent in the way that the failings of the market follow analytically from classical economic theory.
7   The NPM is allied to a broader model in which state interventions should not cause inefficiencies. In this approach the public function of redistribution (i.e. providing welfare to the needy) is increasingly to be separated from the provision of public goods and services. Economic theory has long advocated this, so that redistribution does not alter relative prices and thereby affect private allocation decisions. Assets – ideally money, in theory – are publicly redistributed to people in need (the social 'safety net'), and, exceptionally, to deserving firms (e.g. temporary cash support for an infant industry, rather than tariff protection).
8   An imprest account is a buffer fund of cash from which a department or other unit pays incidental expenses, topped up periodically by payments from central funds.

2. Reforming the Role of Government in Agricultural Markets

1   This chapter is based on Hubbard and Smith (1996).
2   A market fails when it does not encourage efficient production, distribution and consumption of the good or service.
3   Essentials have inelastic demand relative to price.
4   A number of elements determine whether agricultural markets function efficiently (Smith and Ellis 1997). These include the circumstances surrounding the physical transformation of products and the information and financial flows which support this process. Agricultural markets may be inefficient because increasing returns to scale constitute barriers to entry and lead to monopoly in storage, processing and transport activities. Even where there are no obvious 'natural' monopolies, *de facto* monopolies may arise because of the transactions costs incurred in market exchange. These include the identification of trading opportunities, contract negotiation, and monitoring and enforcement (Staatz 1992). The existence of transaction costs may lead to non-competitive behaviour because asymmetric information may confer bargaining advantages on some agents, and because economies of scale are important in transaction costs thereby favouring trading in large quantities. Imperfections in financial markets may mean that those with access to capital can control markets in which they operate. For example, via such interlinked markets, farmers may find themselves forced to reduce prices accepted from

wholesalers because the wholesalers are their only source of credit. These are examined in subsequent chapters in the context of the case-study countries.

5   Thus, the Asians in East Africa, the Lebanese in West Africa, and the Chinese in Asia have all been regarded as 'alien' groups. For example, see Lele and Christiansen 1989.

6   Agricultural policy became understood as a wider matter when research in the 1980s demonstrated that macroeconomic policy (particularly the extent to which currencies deviated from their market value) determined the extent of effective (explicit plus implicit) taxes and subsidies on agriculture more greatly than did direct interventions under agricultural policy (Krueger, Schiff and Valdes 1991).

7   These figures are the median value of transfers in the years 1982–5 for Mexico; 1984–5 for India; and 1988 for China.

8   Marketing boards have been subject to detailed scrutiny and a large literature in the 1980s documented their history and performance (Bates 1981; World Bank 1981; Arhin, Hesp, and Van der Laan 1985; Nellis 1986; Jones 1987; Reusse 1987; Hopcraft 1987).

9   Zambia experienced much initial instability, owing less to the precipitate liberalisation of grain trade in 1993 than to the marketing corporation virtually ceasing to trade.

3. India

1   This chapter consists of extracts from Kohli and Smith (1998), edited and updated by Michael Hubbard.

2   'In the four states of Andra Pradesh, Haryana, Punjab and Uttar Pradesh where FCI obtains the bulk of its rice requirement, state levy prices for rice on average amounted to only 60%–70% of market prices during the period 1992/1993 to 1997/1998' (Umali-Deininger and Deininger 2001: 331).

3   Government of India, 'Agricultural Statistics at a Glance'.

4   'The Reserve Bank of India (RBI) issues directives to Scheduled Commercial Banks (banks fulfilling RBI conditions for amount of paid-up capital and conduct of business) from time to time to regulate credit issued against security of agricultural commodities' (Umali-Deininger and Deininger 2001: 332).

5   'India's self-sufficiency in food is widely regarded as its greatest economic achievement since independence', *The Economist*, 2–8 June 2001, p. 14 of 'A Survey of India's Economy'.

6   Government of India, 'Agricultural Statistics at a Glance'.

7   *The Economist*, 2–8 June 2001, p. 14 of 'A Survey of India's Economy'.

8   Roul (2001) provides a detailed analysis of how over-regulation has frustrated market development in Punjab.

9   Notably in the Budget statements in February 2001 and 2002.

4. Sri Lanka

1   This chapter consists of extracts from Smith and Ellis (1997), edited and updated by Michael Hubbard. A version of the original paper was published as F. Ellis, P. Senanayake and M. Smith, 'Food price policy in Sri Lanka', *Food Policy*, 22, 1, 1997, pp. 81–96.

2   This discussion draws heavily on two papers prepared for the World Bank:
    ARTI 1995 and Harrison 1995.
3   Small-scale producers do appear to be at a disadvantage when it comes to sell-
    ing paddy but this is more marked with respect to sales to the PMB itself.
    Anecdotal evidence suggests that the PMB prefers to buy in bulk from assem-
    blers who have collected paddy from farmers. Clearly, it is more convenient
    for PMB staff to handle bulk purchases than to have to deal with myriad
    purchases and the attendant paperwork.
4   The full analysis is contained in Smith and Ellis 1997: 30.
5   In 1994 these funds provided 30 per cent and 40 per cent of budgetary allo-
    cations to the food stamp and rehabilitation programmes, respectively.

5. Ghana

1   This chapter consists of extracts from Shepherd and Onumah (1997), edited
    and updated by Michael Hubbard.
2   Ghanaian data tends to be inconsistent. Different data series do not agree
    with one another, and apart from price data, which are collected from
    markets, other data tend to be estimated, with the exception of cocoa, where
    marketing is through official channels.
3   In 2000 Cocobod consisted of the following linked companies:
  1   *The Cocoa Marketing Company (Ghana) Ltd. (CMC)*
      • responsible for the external marketing of cocoa beans as well as cocoa
        liquor, cocoa butter and cocoa cake, produced by the Cocoa Processing
        Company Ltd.
  2   *The Cocoa Processing Company Ltd. (CPC, PORTEM)*
      • processes raw cocoa beans into semi-finished products, i.e. cocoa butter,
        liquor, cake or powder;
      • manufactures Golden Tree Brand Chocolate, Couverture 'Pebbles' and
        Vitaco Instant Chocolate Drink.
  3   *The Produce Buying Company Ltd. (PBC)*
      • purchases, stores and distributes cocoa designated takeover centres
        agreed by the Cocoa Marketing Company Ltd;
      • competes with the other eleven private licensed buying companies in
        the internal marketing of cocoa.
  4   *The Cocoa Research Institute of Ghana (CRIG)*
      • investigates problems of coca, kola, coffee, sheanut and the tallow tree
        (Pentadesma butyracea) cultivation;
      • develops planting materials for use by farmers e.g. cocoa
        seedlings/clones and coffee clones;
      • conducts research into the development of other products from cocoa
        waste and by-products.
  5   *The Cocoa Services Division (CSD)*
      • promotes productivity and production of cocoa and coffee through
        well-organised extension services to farmers;
      • responsible for the control of pests and diseases of cocoa and coffee;
      • responsible for multiplication of improved planting material.

6 *Quality Control Division (QCD)*
- responsible for inspection, grading and sealing of cocoa, coffee and Sheanut for export;
- responsible for fumigation and storage of cocoa.

Source: Cocobod website, 2000.

4 'The Cocobod which we inherited [in the early 1980s] was a bloated monster which fed itself on the sweat of our farmers, leaving little for producer prices or revenue for government. We turned it into a downsized, efficient organisation ... We are not against privatisation per se, as we have demonstrated in our divestiture programme ... Cocobod will change, but let us not be in too much of a hurry', President Jerry Rawlings, 2000. FT.com 2000 'Interview with the FT: the journey to democracy', http://specials.ft.com.

5 FT-Com 2000, 'Cocoa: a hot political issue'. Interview with Chairman of Cocobod, http://specials.ft.com.

6 Ghana's cocoa, produced by smallholders, has a higher cocoa butter content, enters into futures trading because of Cocobod's reliability, and commands a premium over other cocoas.

7 Doubts on how efficiently the Ghanaian maize market functions were cast by a study correlating monthly prices between different areas, from the mid 1970s to mid-1980s, which concluded that spatial integration was weak (Sarris and Shams 1991: 144). The same study analysed inter-seasonal price variation and concluded that the profitability of inter-seasonal storage was strong, provided finance is available.

8 FAO Production Statistics, on FAO web page.

9 Consultants employed by USAID reckoned that food aid had created no disincentive to local rice production, on the basis of 'usual marketing requirements'. But so inconsistent are production and trade estimates that the whole exercise of calculating impact is of doubtful value.

10 'The objective of the cocoa strategy is to ensure better performance of the industry and to increase farmers' income. Its highlights include: (i) increasing producer prices to at least 60 per cent of the f.o.b. price for the 1999/2000 season and by at least two percentage points in each of the next two years; (ii) reducing the tax on cocoa from the current level of 26 per cent of the f.o.b. price; (iii) allowing qualified licensed buying companies (LBCs), including the Produce Buying Company (PBC), to export at least 30 per cent of their domestic purchases, starting with the 2000/1 crop; (iv) giving LBCs equal access to Cocoa Marketing Board (Cocobod) warehouses and crop financing; and (v) abolishing price discounts on exportable cocoa to domestic processors. As a result, cocoa production is expected to reach 500 000 tons by 2004/5. Cocobod, the cocoa marketing board, will have a regulatory role and will continue to reduce its share of the f.o.b. price. The extension services of the Cocobod and the Ministry of Agriculture will be unified. The Quality Control Division (QCD) will remain a public sector institution. The PBC will be outsourced for divestiture before end-March 1999 and will be offered for sale by end-June 1999' (IMF 1999).

6. Zimbabwe

1 This chapter consists of edited and updated extracts from Hubbard (1999b).
2 T. Hawkins and J. Lamont writing in the *Financial Times*, 18/19 August 2001.
3 This indicator is selected because it is a practical, if very rough, way to estimate whether changes in the range of public services produced are in the direction of better allocative efficiency, and because the challenge facing Zimbabwe's public agricultural services is greater private sector involvement. More strictly, allocative efficiency in public services improves when the extra social benefit of resources put into different public services becomes more equal. But this is complex to measure.
4 The civil service of Zimbabwe grew rapidly in the 1980s and senior positions were increasingly filled by less experienced blacks as whites left the service (Public Service Review Commission 1989: 6–7).
5 Other factors are thought to be the re-established peace and return to agricultural activity of former fighters (Masst 1996: Chapter 11).
6 '[T]he leading 1–5% of farmers have accounted for the bulk of smallholder marketed outputs for major crops' (World Bank 1994a: 18).
7 By allowing competition in buying and by abolishing movement restrictions on maize.

7. Kenya

1 This chapter consists of extracts from Lewa (1998), edited and updated by Michael Hubbard.
2 There is ample evidence that Kenya has in recent years faced an increasing structural maize deficit (Ministry of Agriculture Reports 1996; Lewa 1995;USAID 1996; FEWS in-depth Report 1, 27 June 1996).
3 Evidence available suggests that Kenya had entered into a structural maize deficit by the mid-1990s, owing mainly to population growth but also to income growth and agricultural diversification (MOALDM Reports 1996).
4 Field research (Lewa 1995) indicated that price differences between the various markets visited reflected transfer costs. This is supported by the Ministry of Agriculture Market information data for the period and WFP, showing prices of maize in various markets in the country. Generally speaking, the prices reflect transfer costs and do indicate that liberalisation had improved pricing efficiency in this case.
5 Prices fell to between Kshs. 350 and 400 in the maize surplus-producing districts. The average cost of producing one 90 kg bag in those areas was Kshs. 600 (MOALDM Reports 1996; NCPB Reports 1996).
6 The NCPB had no finances to purchase maize from farmers and so mandating it to buy meant the government had to look for funds for the exercise.
7 Many large-scale millers informed this researcher that the limited liberalisation of export of maize products, including maize meal since 1995, has helped their businesses a lot in that they are developing new markets and are able to dispose of excess maize products in the regional market. This would appear to

suggest that the complete liberalisation of maize imports and exports is the long-term solution to the maize problem as it will encourage development of new as well as forward markets.

8 In general, prices will fluctuate in the range defined by the difference between net export prices and net import prices. The higher transport, storage, handling and other transaction costs are the wider is this range and therefore the less that international trade can serve to stabilise domestic prices. An unstable exchange rate has a similar effect.

9 From interviews with three of the major grain importers, September 1996.

8. Can Food Supplies be Entrusted to the Market?

1 Government played a significant role only in maize marketing, but never handled more than 10 per cent of marketed maize.

2 A *forward contract* is an obligation to purchase or sell at a later date at a price agreed today. *Futures contracts* evolved out of forward contracts. They are forward contracts which are traded on an exchange, thereby enabling the contractors to be relieved of the obligation to fulfil the contract agreed by selling their part in the contract to another party. An *option on a futures contract* gives the buyer the right to buy or sell a futures contract at a later date at a price agreed today (Chance 1995: 4–5).

3 *The Economist*, 2–8 June 2001, 'A Survey of India's Economy'.

4 The research uncovered the following examples of uncertainties for private trade which incomplete privatisation of public marketing agencies can cause: Firstly, the use of the near-moribund Paddy Marketing Board by the Sri Lankan government in the mid-1990s to increase rice buying for political purposes, at the same time as it was using US food aid to reduce imported wheat prices (Smith and Ellis 1997: 40). Secondly, in Zimbabwe in the mid-1990s, the government's use of the formally independent Grain Marketing Board (GMB) to increase maize purchases at subsidised prices, in addition to the abuse by the 'independent' GMB of its continuing monopoly of external trade in maize to raise domestic maize prices in order to increase the value of its own sales (Hubbard 1999: 36)

5 Research for developing the Inventory Credit Scheme to assist maize trade development found that even for wholesale traders, poor book-keeping practices impede information disclosure about their operations, making cash flow analysis in support of loan applications particularly difficult (Shepherd and Onumah 1997: 61).

6 See 'Warehouse Receipts: Financing Agricultural Producers', technical note 5, October 2000, in the series 'Innovations in Microfinance' for more detail on this initiative, at www.mip.org/pdfs/mbp/technical_note-5pdf.

7 Sri Lanka is close to self-sufficiency in rice. As a national food security precaution it bonds licensed rice importers to hold a minimum stock in government warehouses. See Smith and Ellis 1997.

8 By October 2001, news reports indicated that privately owned stocks in GMB stores had been requisitioned at controlled prices. The army was patrolling farms to ensure that sales only took place at GMB prices. The attempt at securing stocks of maize at sub-market prices was not containing market prices,

and appeared to be threatening plantings of maize. In its closing announcement ZIMACE stated: 'in light of the fact that there was absolutely no consultation between the Ministry of Lands and Agriculture and the industry as a whole, it is unclear exactly what the objectives of this new legislation may be ... We have failed to establish any positive effects; there is a shortage of maize in Zimbabwe and the new legislation will not improve the situation, if anything it will deteriorate.'

9   ZIMACE had only operated a forward market, and Kenya's exchange (KACE) does not yet run auctions.

10   Stagnant (Kenya) or shrinking (Zimbabwe) GNP per capita, rapidly decreasing rural per cent of population (84 per cent to 68 per cent Kenya, 78 per cent to 65 per cent Zimbabwe during 1980–99), plus declining human development index during the 1990s, suggest poor performance in rural development. Ghana's performance was little better. Source: World Bank 2000; UNDP 2001.

11   An overview of the variety of approaches and issues regarding safety nets is provided by World Bank 2000.

12   The state food purchase, storage and sale which provided food-based relief were not designed to have high fiscal costs. To operate without state subsidy these interventions needed to pay farmers no more than the selling price less operating costs of the scheme. Operated in that way such schemes put no fiscal burden on governments. But their failure to contain costs, plus political pressures to pay higher prices to famers (e.g. pan-territorial pricing in Kenya and Zimbabwe) made them a heavy fiscal burden, and is speeding their demise.

9. Can Public Marketing Agencies be Reformed?

1   Public components are those which would not be provided (financed and delivered) by private firms. See further discussion in Chapter 1.

2   This section is based on Hubbard 1999.

3   See Jackson 2002 for further analysis of structural adjustment and the textile sector in Zimbabwe.

4   CSC's monopoly of meat sales to the cities was at the expense of city consumers, since smaller, less formal abattoirs (such as those serving the small towns) could supply meat much more cheaply than could CSC. As had occurred in Kenya in the 1970s, pressure mounted for liberalisation of meat movement and for permits to be granted for private abattoirs in the cities. In 1993 trade restrictions were abolished, and CSC (like Kenya Meat Commission in the 1970s) dramatically lost market share. Meat market liberalisation in South Africa has had similar results for the large abattoirs.

5   As a GMB representative put it in an interview: 'we get orders from the Ministry of Agriculture setting the prices we must buy or sell at. But the Ministry of Finance refuses to pay us for these non-commercial activities.'

6   The main source for grain product imports would be South Africa. Since the South African market is much larger than Zimbabwe's, an active and open cross-border market would tend to stabilise prices at South African levels, by quality.

7 It was pointed out by GMB that it was not worth their while bringing stocks from their depots to sell in cities when the selling price was fixed below the market price.

8 In 2000 the GMB's director, together with the Minister and Permanent Secretary of the Ministry of Agriculture, were charged with embezzlement via external grain trading. At the time of writing (2002) the charges were still pending.

9 See Chapter 8 above for further discussion of ZIMACE.

10 This section is based on Smith and Ellis 1997.

11 Paddy farmers protest that they have been unable to sell their harvest for a price that covers their cost of production. In August 2000 paddy farmers in Polonnaruwa launched a 7000-strong demonstration in Hingurakgoda. It had support from the sangha, with three monks joining eight farmers in a fast-unto-death. It was called off when government acceded to their main demand with an assurance that it would buy the sudu nadu variety of paddy for Rs.13 a kilo. The twenty-nine demands placed before the government by farmers include the introduction of policies that would allow them to set the selling prices for their products; an immediate stop to imports of domestic produce such as rice, potatoes, onions and chillies; and the call for continued government support in matters such as marketing, provision of credit facilities, agricultural extension services, distribution of good quality seed paddy and revival and maintenance of ancient tanks (*Sunday Times Sri Lanka* on line, 29 October 2000, 'For a few rupees more: killer agricultural policies take their toll').

12 This section is based on Shepherd and Onumah 1997.

13 *www.FTcom-Ghana*, 'Cocoa: a hot political issue', 2000.

14 FT.Com 2000, 'Cocoa: a hot political issue', http://specials.ft.com.

15 The EU decision to allow a higher proportion of vegetable fat in chocolate contributed to recent weakening of the market.

16 Marketing costs in Ghana were 17 per cent of f.o.b. prices compared to 10–12 per cent elsewhere (LMC International 1996a).

17 The public took up less than the 4 per cent offered to them. 'Members of the Ghana Farmers' Association bought just over 1% of the 20% made available to them. Former and current employees of the PBC took up the 5% reserved for them. Private foreign investors willing to take up about 30% are still sought', *Oxford Analytica*, 15 Nov. 2000, 'Ghana – Vulnerable Economy', www.oxan.com/columns/wkcol_15112000.html.

18 Africa Online, Cocoa Commodity News 1999, 'Cocoa processing in Ghana', http://www.otal.com.

19 The public role in quality assurance, and the case of QCD, is discussed in more detail in Chapter 11.

20 Gilbert, Varangis and ter Wengel (1999) suggest that public marketing agencies for coffee might usefully be converted into organisations to lobby for the industry's interests with government.

21 This section is based on Lewa 1998.

22 'The impending bumper maize harvest in Kenya's "grain basket" is bad news for farmers … The bad news is that the country is awash with maize at a time when national stocks are traditionally low. Many farmers are holding sizeable stocks estimated to be in the region of 200,000 metric tonnes. Many of the silos of NCPB are brimming with last season's harvest which it is

having a problem selling. This is not really surprising since NCPB bought a bag at Sh1,250 and at present, maize is readily available on the market at Sh800 to Sh900 a bag and the price is certain to go down as the harvest gets into full throttle' (Robert Shaw, *The Nation* (Nairobi), 7 October 2001).

23   No official estimate exists of how much NCPB costs are raised by corruption and waste. An anonymous paper circulated in the mid-1990s estimated that fraud and waste accounted for over 60 per cent of NCPB's costs.

24   The EC pulled out its support, as did other donors, to force government to adopt a multi-party system of government. Training of staff and support for the private sector suffered a similar fate. Government retaliated by questioning how serious donors were regarding reform.

25   The significance of functioning foreign exchange markets for national food security policy should not be underestimated: out of a sample of thirty African countries during the period 1965–83, twenty-seven experienced foreign exchange shortage as the binding constraint preventing food imports from being used to ensure national food availability (Kirkpatrick and Diakosavvas 1985).

26   CSRP should have contained more direct assistance to private sector development 'Kenya: Country Assistance Memorandum', World Bank, 20 November 2000.

27   Infrastructure and public services have deteriorated, growth rates have decreased, and there is widespread corruption in state corporations, according to a variety of sources, e.g. *Africa Confidential*, vol. 42, 18, 14 September 2001, 'Moi vs the economy'.

28   'The fact of the matter is that Kenyan maize, and indeed some of our other food products, are expensive by both regional and international yardsticks. Wholesale maize prices in Uganda and Tanzania are around half and two thirds respectively ... those in Kenya. What is evident is that something is clearly wrong on the production side and input side. It is easy, but unhelpful, for Kenyan farmers to blame millers, the government or some other farmer in another country whose cost of production is lower than here'. 'Exorbitant prices hurt trade, consumers', Robert Shaw, *The Nation*, Nairobi, 7 October 2001.

29   This section is based on Kohli and Smith 1998.

30   'Indonesia's farmers in uproar over cheap rice imports', *Asia Times* on line, 10 March 2000.

31   Bangladesh's grain policy reform has focused on raising production and productivity, bringing domestic prices in line with external prices, and using resources from food aid to bring political forces behind abolition of the universal ration in favour of targeting (Ahmed *et al.* 2000). The reform effort does not appear to have prioritised reform of the public grain marketing agency.

10. Can Public Services to Marketing be Contracted Out?

1   Prices may be more certain, there are no competitors whose behaviour needs to be taken into account, while government is, in theory at least, in a strong position to enforce contracts, through its resources and personnel.

2   This section is based on Smith and Ellis 1997.

3   Evidence that this arrangement is working well is provided by the fact that when the researchers visited the FCD on 24 January 1996 the 'Stock Position' data for the companies as of 22 January 1996 were already available.

4   The Cooperative Wholesale Establishment is a public organisation established in 1949. It was originally expected to purchase and distribute a range of consumer goods, including some food items. In 1989 the CWE took over the monopoly of wheat imports from the Food Commissioner's Department.

5   This section is based on Lewa 1998.

6   One miller pointed out that three powerful politicians had shares in two of the firms that won the tenders to export maize. These are the same firms that did not execute their tenders by the end of the contract period but had no problem with extending their contracts.

7   Government in 1995 took the further step of creating a maize exports fund into which the proceeds of exports were paid. Under the export sales contracts, payments for maize exports were made on the basis of 90 per cent upfront payment per ship lot after presentation of all the relevant export documents, including evidence that the maize was in a warehouse in Mombasa. This enabled maize farmers to be paid in good time. Previously farmers suffered very long payment delays as a result of NCPB procedures.

8   During the loading at the port of Mombasa in October 1995 for the export of the second contract, the researcher observed that while the loading of maize was previously 1200mt per day on average, the private sector was able to raise the figure to 2500mt per day.

9   The only Kenya railways vessel plying Lake Victoria for the maize contract, *MV Uhuru*, was grounded due to the expiry of its insurance. The expiry was a shock because exporters woke up one morning only to be told that the vessel did not have insurance. At the time sixty-three wagons were already loaded and waiting in Kisumu. There was no other vessel to handle the kind of volume that was expected to be shipped.

10   Interview with NCPB's General Manager Operations in Mombasa during shipment.

11   The exporters had complained to the Ministry of Agriculture that NCPB was not cooperating (MOALDM Reports 1996). The Board retorted that the complaints were not justified. A visit by this researcher to Kisumu in October 1996 provided evidence that there was a problem of coordination between NCPB, the exporters and the firms receiving grain on the Tanzania side.

12   This section is based on Kohli and Smith 1998.

13   In 2000 a National Policy on Handling, Storage and Transportation of Foodgrains was announced 'In order to reduce storage and transit losses of foodgrains at farm and commercial level, to modernise the system of handling, storage and transportation of the foodgrains procured by the Food Corporation of India (FCI) and to bring in additionality of resources through private sector involvement.'

14   Spot contracts are six to eight months or more. Long-lease contracts consist of a few years' guarantee by the FCI for use of the facility, assuring the agent of rent for this period.

15 There are two types of storage facilities that FCI owns/hires: covered
    go-downs and CAP (covered and plinth). CAP storage facilities are open,
    i.e. without concrete cover but with polythene covers, and may or may not
    be walled around. It is mainly CAP storage which is associated with heavy
    storage losses and which is now targeted for replacement.

16 Most grain stores even in the agrarian states of Punjab, Haryana and Uttar
    Pradesh have been constructed under guarantees and with public financial
    assistance, suggesting that even in areas of high demand, investment in stor-
    age facilities has had to be induced by the state. The chief reason (according
    to interviews with private entrepreneurs) seems to be the heavy capital
    outlay, particularly for large stores of 5000 tonnes. Smaller stores (e.g. 1000
    tonnes) are mostly industrial warehouses and owned by millers and
    wholesale traders. These are more easily financed and have lower risks owing
    to their multi-purpose use.

17 The Agriculture Refinance and Development Corporation (ARDC) was the
    predecessor of NABARD (National Bank for Agriculture and Rural
    Development), the premier development financial institution in India for
    agricultural finance.

18 Liberalisation of the Indian economy has opened the way for private invest-
    ment in container facilities at ports. Major shipping companies, transport
    and stevedoring companies and shipping agents have entered the sector, as
    well as foreign logistics and warehousing companies, by starting joint ven-
    tures with Indian partners. These now offer competition to market leaders
    like CWC.

19 This section is based on Shepherd and Onumah (1997).

11. How can Quality be Assured?

1 Hazardous Action Critical Control Points (HACCP) identifies points in the
   production process that are most critical to monitor and control (Henson
   and Caswell 1999). It is a process-based approach which allows for more
   flexibility (e.g. in facilities required for processing, and thereby potentially
   more effective and lower cost quality control by management.

2 This section is based on Kohli and Smith (1998).

3 These include APEDA (Agricultural Products Export Development
   Authority), the Spices Board of India (spices only), and the Food Products
   Office, which is attached to the Ministry of Food and quality control
   activity is restricted to fruit juices, jams, and canned food products. The
   activities of these organisations often extend well beyond quality control
   encompassing promotional efforts to develop new markets for their
   products.

4 A more credible, well-known and active role in export certification is played
   by the BIS. The export certification BIS label ISO 9000 is internationally
   known and understood by overseas buyers and offers global entry to Indian
   exporters. Consequently, it is much sought after by exporters for its international credibility.

5 This section is based on Shepherd and Onumah 1997.

6   Since privatisation of internal marketing in 1991 buyers are the licensed buy-
    ing companies (LBCs) as well as Cocobod's Produce Buying Company. See
    Chapters 5 and 9 for further detail.
7   The PPRC is composed of Cocobod officials, representatives of farmers, LBCs
    and government.
8   Privatisation of the domestic trade in sheanuts was argued by some to lead
    to farmers compromising on boiling and drying of nuts to reduce moisture
    content from 45 per cent to 8 per cent, with the result that free fatty acid
    content had increased. A similar decline in quality was argued to have
    occurred in Nigeria and Benin with privatisation of sheanut trade.
9   The allegations were that QCD's grading system for coffee beans was
    different from internationally accepted standards; tests using meshes to get
    a regular size of bean were not applied, with the result that Ghanaian cof-
    fee sells at a discount. Sheanuts are inspected in Ghana by looks, whereas
    the important quality criterion is the proportion of free fatty acid (FFA).
    QCD argued that no decline in the quality of sheanuts exported had been
    recorded by QCD since liberalisation of the domestic buying in the early
    1990s. Coffee quality was perceived to have declined since domestic
    liberalisation as a result of increased demand and better prices, which
    encouraged producers and traders to add broken beans to their con-
    signments. The QCD has tried to ensure that such consignments go direct
    to processors, and do not enter the free market where they could damage
    the country's reputation.
10 This section is based on Atse 1999.
11 This section is based on Sukume 1999.
12 'The regulatory functions of the Dairy Services and Meat Grading Services
    with the Livestock and Pastures Division have been severely jeopardised by
    lack of transport and limited subsistence and travel provisions. The
    Livestock and Pastures Division is responsible for the grading of all carcases
    from all animals slaughtered in registered abattoirs throughout the country,
    to monitor regularly compliance with grading regulations in abattoirs where
    approved non-government graders are used, and to grade all live animals for
    sale on public auctions in communal areas. It is also responsible for the
    inspection and registration of dairy units and for regular monitoring of milk
    quality ... These activities require regular visits throughout the country.
    Without regular monitoring, quality standards can be expected to continue
    to deteriorate' (Tawonezvi and Mapika 1995: 8).
13 Provisions were made by the Ministry of Finance in the late 1990s allowing
    departments to retain revenues in revolving funds.
14 This section is based on Rusike 1999.

12. What Public Role is There in Market Information?

1   This section is based on Shepherd and Onumah 1997.
2   This section is based on Kohli and Smith 1998.
3   The major publications of the DES include: *Agricultural Situation in India*
    (monthly journal), the *Bulletin on Agricultural Prices* (weekly), *Indian Agriculture
    in Brief, Area and Production of Principal Crops in India* (annual), *Agricultural*

Prices in India (annual), *Agricultural Wages in India* (annual), *Indian Agricultural Statistics* (annual), *Farm Harvest Prices of Principal Crops in India* (annual) and *Government of India, Agricultural Statistics at a Glance* (annual). All these publications are bilingual, i.e. in English and Hindi. In addition, two other Hindi publications are also published by the Directorate.

4   This section is based on Smith and Ellis 1997.

5   This section is based on Chikandi 1999.

6   Farmers' associations do successfully run market information services for their members – but these tend to be more specialised organisations (e.g. coffee exporters) with relatively few large members (Gilbert, Varangis and ter Wengel, 1999).

7   www.fao.org/ag/magazine/0104sp.htm.

13. Developing Agricultural Trade

1   Facilitating trade development involves both deepening of trade (e.g. emergence of futures and options markets) and widening of trade (enabling underdeveloped areas to participate in trade).

2   This chapter is based in part on Hubbard and Smith 1999.

3   The growth in private trade within and between countries is only partly the result of removal of government controls on trade. Over the long term it appears to be associated with increasing population density, higher incomes and better transport and communication.

4   Jayne, Govereh *et al.* (2001) detail reversals in trade liberalisation in Ethiopia, Kenya and Zambia, particularly in the fertiliser trade. In Zimbabwe private trading in maize was suspended in 2001.

5   A corporatised public agency is one which remains publicly owned but manages its own business under the direction of an independent board.

6   Competition for public work is the logical extension of corporatisation under the 'new public management', as in the UK with compulsory competitive tendering for local government services, or use of 'best value' comparisons between public and private providers of a service.

7   Government's capacity is what determines its performance (see Chapter 1). High capacity exists where government is able to make good use of its human and material resources to achieve its purpose. To do so requires political authority, ability to attract additional resources (e.g. from donors) and ability to make and implement policy.

8   Government announced a package of reforms in February 2002, including lifting of restrictions on trade and storage of grains. But similar reforms have been announced in the past (e.g. budget statement 2001), therefore substantial reform may be longer in coming.

References

Ahluwalia, M. (1996) 'New economic policy and agriculture: some reflections', *Indian Journal of Agricultural Economics*, 51: 3, 412–26.

Ahmed, R., Haggblade, S. and Chowdhury, T. (2000) 'Out of the shadow of famine: evolving food markets and food policy in Bangladesh', IFPRI Food Policy Statement, No. 31, August (Washington: International Food Policy Research Institute).

Arhin, K., Hesp, P. and Van der Laan L. eds (1985) *Marketing Boards in Tropical Africa* (London: Kegan Paul International).

ARTI (1995) 'Study of domestic agricultural marketing in Sri Lanka', Colombo.

Atse, D. (1999) 'Reforming the cocoa sector in Côte d'Ivoire: the case of the electronic trading system', Paper 34, The Role of Government in Adjusting Economies University of Birmingham, School of Public Policy.

Attwood, E.A. 'Comparative efficiency in the dairy sectors of Zimbabwe and New Zealand and the effects of the 1994 GATT agreement on the development of the dairy sector of Zimbabwe' (Volume 1: Marketing Systems), mimeo, no date.

Barrett, C. (1998) 'Immiserized growth in liberalized agriculture', *World Development*, 26: 5, 743–53, May (Elsevier).

Bates, R. (1981) *Markets and States in Tropical Africa: the Political Basis of Agricultural Policies* (California: University of California Press).

Bates, R. (1989) 'Structural adjustment and agriculture', in S. Commander (ed.), *Structural Adjustment and Agriculture: Theory and Practice in Africa and Latin America* (London: ODI).

Batley, R. (1997) 'A research framework for analysing capacity to undertake the "new roles" of government', Paper 23, The Role of Government in Adjusting Economies. Birmingham: School of Public Policy.

Bayley, B. (2000) *A Revolution in the Market: the Deregulation of South African Agriculture* (Oxford: Oxford Policy Management).

Belshaw, D., Lawrence, P. and Hubbard, M. (1999) 'Agricultural tradables and economic recovery in Uganda: the limitations of structural adjustment in practice', *World Development*, 27: 4, 673–90, April.

Bhalla, G. (1994) 'Policy for food security in India', in G. Bhalla (ed.), *Economic Liberalization and Indian Agriculture* (New Delhi: Institute for Studies in Industrial Development).

Booker and Githongo Associates (1988) 'Grain Marketing Study', Nairobi.

Borsdorf, R. (1993) *Cereal consumption in Sri Lanka* (Colombo: USAID).

Buchanan, J. (1986) *Liberty, market and state* (Brighton: Wheatsheaf).

Chance, D. (1995) *An Introduction to Derivatives* (Orlando: Dryden Press).

Chikandi, S. (1999) 'ZFU's pilot market information service: an analysis of the service and early consumer reaction to it', in M. Hubbard (ed.), *The Impact of liberalisation on public services to agricultural marketing: case studies of change options in Zimbabwe*. The Role of Government in Adjusting Economies, Paper 36. University of Birmingham, School of Public Policy.

Chilowa, W. (1998) 'The impact of agricultural liberalisation on food security in Malawi', *Food Policy*, 23: 6, 553–69, December (Elsevier).

Coase, R. (1937) 'The nature of the firm', *Economica* (new series), 9, 386–405.

Collier, P. *et al.* (1995) *Zimbabwe: Consolidating the Trade Liberalisation* (UNDP and World Bank, trade expansion programme).

Commander, S., Howell, J. and Seini, W. (1989) 'Ghana 1987–93', in S. Commander (ed.), *Structural Adjustment and Agriculture: Theory and Practice in Africa and Latin America* (London: ODI).

Coulter, Jonathan (1993a) 'The Malian experience in financing the grain trade', *African Review of Money Finance and Banking*, 1, 27–45.

Coulter, J. (1993b) 'Grain market liberalization in Sub-Saharan Africa: lessons in practical implementation and promotion of private sector development', Paper presented at NRI–IFPRI Research Symposium on Critical Food Policy Issues for Sub-Saharan Africa, 24–25 March 1993.

Coulter, J. and Asante, E. (1993) *New Approaches to the Financing of Agriculture in Ghana Natural Resources Institute* (London: NRI and ODA).

Coulter, J. and Onumah, G. (2001) 'Enhancing rural livelihoods through improved agricultural commodity marketing in Africa: role of warehouse receipts'. Paper presented at 74th EAAE Seminar on Livelihoods and Rural Poverty: Technology, Policy and Institutions. Imperial College at Wye, UK.

Coulter, J. and Shepherd, A. (1995) *Inventory Credit: an Approach to Developing Agricultural Markets*, FAO Agricultural Services Bulletin No. 120, FAO, Rome.

DAI and IDA (1989) *Economic and Social Soundness Analyses for the Kenya Market Development Programme* (Nairobi: USAID).

Danida (1993) *Evaluation of Danish Assistance to Grain Storage and Grain Drying Projects, Vol. II* (ref. no. Dan.4/52–16 Eval.), Danida/Ministry of Foreign Affairs.

Dapaah, S. K. (1993) 'Overview of the structure of agriculture in Ghana', Paper presented at workshop on financing microagricultural enterprises in Ghana at GIMPA, Accra, February, 1993.

Dorosh, P. (2001) 'Trade liberalization and national food security: rice trade between Bangladesh and India', *World Development*, 29: 4, April, 673–89.

Edwards, C. (1993) *Protectionism and Trade Policy in Manufacturing and Agriculture: Sri Lanka* (Colombo: Institute of Policy Studies).

Eicher, C. (1995) 'Zimbabwe's maize-based green revolution: preconditions for replication', *World Development*, 23: 5, 805–17.

Ellis, F. (1993) 'Private trade and public role in staple food marketing: the case of rice in Indonesia', *Food Policy*, 18: 5, 428–38.

Ellis, F. P. Senanayake and M. Smith (1997) 'Food price policy in Sri Lanka', *Food Policy*, 22: 1, 81–96.

FAO (1991) 'Agricultural price and marketing policy: government and the market in Africa'. Policy Analysis Division, Training Service (Espt) (Rome: FAO).

Famine Early Warning System In-depth Report 1 (USAID), 27 June 1996.

Findlay, Christopher *et al.* (1993) *Policy Reform, Economic Growth and China's Agriculture* (Paris, OECD).

Gilbert, C., Varangis, P. and ter Wengel, J. (1999) 'Industry organisations and market reforms in selected coffee producing countries'. Report of a study undertaken for World Bank and Common Fund for Commodities. Available on World Bank website.

Goletti, F. (1994) 'The changing public role in a rice economy approaching self-sufficiency: the case of Bangladesh', IFPRI Abstract, research report 98, September (Washington: International Food Policy Research Institute).

Government of India (1996) *Ministry of Civil Supplies: Annual Report.*

Government of Zambia (1993) *A Review of Crop Marketing Liberalization in 1993: the Need for a Transition Programme* (Lusaka: MAFF).

Haggblade, S., Hazell, P. and Brown, J. (1989) 'Farm-nonfarm linkages in rural sub-saharan Africa', *World Development*, 17: 8, 1173–1201.

Harden, I. (1992) *The Contracting State* (Buckingham: Open University Press).

Harrison, P. (1995) 'Domestic marketing in the non-plantation crops sector of Sri Lanka', Working paper no. 1 (Colombo: World Bank).

Hazell, Peter and Rigoberto, Stewart (1993) 'Should Costa Rica's grain markets be liberalised?', *Food Policy*, 18: 6, 471–81.

Henson, S. and Caswell, J. (1999) 'Food safety regulation: an overview of contemporary issues', *Food Policy*, 24: 589–603.

Hesselmark, O. and Lorenzil, G. (1976), 'Structure and performance of the maize marketing system in Kenya', *Zeitschrift fuer Auslaendische Landwirtschaft*, 15 (2): 171–9.

Hicks, J. (1969) *A Theory of Economic History* (Oxford: Oxford University Press).

Hine, J. L. and Riverson, J. D. N. (1982) 'The impact of feeder road investment on accessibility and agricultural development in Ghana', in *Highway Investment in Developing Countries* (London: Thomas Telford Ltd.).

Hopcraft, P. (1987) 'Grain marketing policies and institutions in Africa', *Finance and Development*, 24: 1 (Washington).

Hubbard, M. (1988) 'Drought relief and drought-proofing in the state of Gujarat, India', in D. Curtis, M. Hubbard and A. Shepherd, *Preventing Famine: Policies and Prospects for Africa* (London: Routledge).

Hubbard, M. (1997a) 'The new institutional economics in agricultural development: insights and challenges', *Journal of Agricultural Economics*, 48: 2, 239–49, May (UK: Agricultural Economics Society).

Hubbard, M. (1997b) 'Cost recovery, commercialisation, outsourcing and decentralisation of agricultural services in Zimbabwe: opportunities and issues', Annex note. Ministry of Agriculture Zimbabwe, Agricultural services and Management Project Pre-appraisal.

Hubbard, M. (ed.) (1999a) 'The impact of liberalisation on public services to agricultural marketing: case studies of change options in Zimbabwe', Paper 36, The Role of Government in Adjusting Economies, University of Birmingham.

Hubbard, M. (1999b) 'Governing open agricultural markets in Zimbabwe: state capacity and performance', Paper 37, The Role of Government in Adjusting Economies, University of Birmingham, School of Public Policy.

Hubbard, M. and Smith, M. (1996) 'Agricultural marketing sector review', Paper 6, The Role of Government in Adjusting Economies, University of Birmingham, School of Public Policy.

Hubbard, M. and Smith, M. (1999) 'Does New Public Management work in reforming the state's role in agricultural marketing in developing countries ?' *Journal of International Development*, 11: 785–90.

IMF (1999) *Ghana: Enhanced Structural Adjustment Facility Policy Framework Paper 1999–2001* (Washington: IMF).

ISSER (1995) *The State of the Ghanaian Economy in 1994* (Accra: University of Ghana: Institute of Statistical, Social and Economic Research).

ISSER (1996) *The State of the Ghanaian Economy in 1995* (Accra: University of Ghana, Legon).

Jackson, P. (2002) *Business Development in Asia and Africa: The Role of Government Agencies* (Basingstoke: Palgrave).

Jaffee, S. and Morton, J. (eds) (1995) *Marketing Africa's High Value Foods: Comparative Experiences of an Emergent Private Sector* (Iowa: Kendall/Hunt).

Jayne, T. and Jones, S. (1997) 'Food marketing and pricing policy in eastern and southern Africa: a survey', *World Development*, 25: 9, 1505–27.

Jayne, T., Govereh, J., Mwanaumo, A., Chapoto, A. and Nyoro, J. (2001) 'False promise or false premise? The experience of food and input market reform in eastern and southern Africa', paper to EAAE seminar on livelihoods and rural poverty, Imperial College at Wye, September.

Jayne, T., Rubey, L., Chisvo, M. and Weber, M. (2001) 'A food security success story: maize market reforms improve access to food even while government eliminates food subsidies', Michigan State University, Agricultural Economics Working Papers.

Jones, Stephen (1994) *Agricultural Marketing in Africa: Privatisation and Policy Reform* (Oxford, Food Studies Group).

Jones, Stephen and Beynon, Jonathan (1992) 'Market reform and private trade in Eastern and Southern Africa', *Food Policy*, 17: 6.

Jones, W. (1987) 'Food crop marketing boards in tropical Africa', *Journal of Modern African Studies*, 25: 3, 375–402.

Joshi, V. and Little, I. (1994) *India: Macroeconomics and Political Economy 1964–1991* (New Delhi: OUP).

Joshi, V. and Little, I. (1996) *India's Economic Reforms 1991–2001* (New Delhi: OUP).

Kirkpatrick, C. and Diakosavvas, D. (1985) 'Food insecurity and the foreign exchange constraint in developing countries', *Journal of Modern African Studies*, 23: 2, June.

Knudsen, Odin *et al.* (1988) 'Redefining the role of government in agriculture for the 1990s', World Bank Discussion Paper, no. 105 (Washington D.C.: World Bank).

Kofi, T. (1993) *Structural Adjustment in Africa: a Performance Review of World Bank Policies under Uncertainty in Commodity Price Trends: the Case of Ghana* (Helsinki: United Nations University World Institute for Development Economic Research).

Kohli, R. and Smith, M. (1998) 'The role of the state and agricultural marketing reform in India', Paper 29, The Role of Government in Adjusting Economies, University of Birmingham, School of Public Policy.

Kotey, R. *et al.* (1974) 'The economics of cocoa production and marketing', Accra: University of Ghana, Institute of Statistical, Social and Economic Research.

Krishnaji, N. (1991) 'Agricultural price policy in Asia', *Indian Journal of Agricultural Economics*, 46: 2, 186–92.

Krueger, A., Schiff, M. and Valdes, A. (1991) *The Political Economy of Agricultural Pricing Policy: Volume 3, Africa and the Mediterranean* (Baltimore: Johns Hopkins University Press, for the World Bank (Volumes 1 and 2 cover Latin America and Asia).

Larbi, G. (1995) *Implications and Impact of Structural Adjustment on the Civil Service: the Case of Ghana – Paper 2*, The Role of Government in Adjusting

Economies, Development Administration Group, University of Birmingham, Birmingham.

Larson, D., Varangis, P. and Yabuki, N. (1998) 'Commodity risk management and development'. Paper prepared for the Roundtable Discussion on New Approaches to Commodity Price Risk Management in Developing Countries (Washington: World Bank).

Lele, Uma and Christiansen, Robert E. (1989) 'Markets, marketing boards, and adjustment policy in Africa: issues in adjustment policy', MADIA Discussion Paper, no. 11 (Washington D.C.: World Bank).

Lewa, P. (1995), 'The politics of grain market reform in Kenya: a study of the maize milling industry'. Doctoral thesis, University of Birmingham, School of Public Policy.

Lewa, P. (1998) 'Managing the public role in a liberalised maize marketing environment in Kenya', Paper 30, The Role of Government in Adjusting Economies University of Birmingham, School of Public Policy.

Lewa, P. and Hubbard, M. (1995), 'Kenya's cereal sector reform programme: managing the politics of reform', *Food Policy*, 20: 6, 573–84.

LMC International (1996a) *The External Marketing of Ghana's Cocoa* (Accra: Ministry of Finance).

LMC International (1996b) *Commodity Bulletin: Cocoa* (Oxford: LMC International), July.

Malkin, J. and Wildavsky, A. (1991) 'Why the traditional distinction between public and private goods should be abandoned', *Journal of Theoretical Politics*, 3: 4, 355–78.

Masst, M. (1996) 'The harvest of independence: commodity boom and socio-economic differentiation among peasants in Zimbabwe', PhD dissertation, University of Oslo.

Mather, G. (1989) 'Thatcherism and local government: an evaluation', in J. Stewart and G. Stoker (eds), *The Future of Local Government* (London: Macmillan).

Menegay, M. *et al.* (1995) *Assessment of Domestic Wholesale Markets: Marketing of Fresh Vegetables in Sri Lanka* (Colombo: USAID).

Ministry of Agriculture, *Livestock Development and Marketing Reports*, 1992; 1995; 1996 (Nairobi).

Minot, N. and Goletti, F. (2001) 'Rice market liberalization and poverty in Vietnam', Research report 114 (Washington: IFPRI).

Mooij, J. (1995) 'The political economy of the Essential Commodities Act', Development Policy and Practice Research Group, Open University, DPP Working Paper No. 29.

Nellis, J. (1986) 'Public enterprises in Africa', World Bank Discussion Papers (Washington).

Nindi, Benson C. (1990) 'Agricultural marketing reforms and the public vs. private debate in Tanzania', *Journal of Rural Cooperation*, 18: 1, 3–29.

North, D. (1990) *Institutions, Institutional Change and Economic Performance* (Cambridge: Cambridge University Press).

Nyanteng, V. (1980) 'The declining Ghana cocoa industry: an analysis of some fundamental problems' (Accra: University of Ghana, Institute of Statistical, Social and Economic Research).

Nyanteng, V. (1995) 'A review of the internal marketing of cocoa following the introduction of competition in 1992'. Report on a study commissioned by the Ghana Cocoa Board.

Nyanteng V. K. and Dapaah, S. K. (1993) *Agricultural Development in Policies and Options for Ghanaian Economic Development* (ed. V. K. Nyanteng) ISSER.

Ofosu, S. (1995) *Ghana's Cocoa Rehabilitation Project in the Context of the Economic Recovery Programme/Structural Adjustment Programme: How Successful?* (Tokyo: Institute of Developing Economies).

Ofreneo, René, E. (1987) 'Deregulation and the agrarian crisis' (Quezon City: University of the Philippines).

Owango, M., Staal, S., Kenyanjui, M., Lukuyu, B., Njubi, D. and Thorpe, W. (1998) 'Dairy cooperatives and policy reform in Kenya: effect of livestock service and milk market liberalisation', *Food Policy*, 23: 2, 173–85.

Palaskas, T. and Harriss-White, B. (1993) 'Testing market integration: new approaches with case material from West Bengal', *Journal of Development Studies*, 30: 1, 1–57.

Pletcher, J. (2000) 'The politics of liberalizing Zambia's maize markets', *World Development*, 28· 1, 129–42 (Elsevier).

Public Service Review Commission (1989) 'Report of the Public Service Review Commission of Zimbabwe (Kavran Commission)', May (Harare: Government printer).

Pursell, G. and Gulati, A. (1995) 'Liberalizing Indian agriculture: an agenda for reform', in R. Cassen and V. Joshi (eds), *India: the Future of Economic Reform* (Delhi: OUP).

Radakrishna, I. (1988) 'India', in *Evaluated Rice Market Intervention Policies: Asian Examples* (Manila: Asian Development Bank).

Reardon, T. (1997) 'Using evidence of household income diversification to inform study of the rural nonfarm labor market in Africa', *World Development*, 25: 5, 735–47.

Reardon, T., Berdegue, J. and Escobar, G. (2001) 'Rural nonfarm employment and incomes in Latin America: overview and policy implications', *World Development*, 29: 3, 395–409 (special issue).

Reusse, R. (ed.) (1987) 'Liberalization and agricultural marketing: recent causes and effects in third world economies', *Food Policy*, November.

Roul, C. (2001) *Bitter to Better Harvest: Agricultural and Marketing Strategy for India* (New Delhi: Northern Book Centre).

Ruf, F. (1995) *Booms et crises du cacao: Les vertiges de l'or brun* (Paris: Karthala).

Rusike, J. (1999) 'Managing the public role in liberalised seed markets in Zimbabwe', in M. Hubbard (ed.) *The Impact of Liberalisation on Public Services to Agricultural Marketing: Case Studies of Change Options in Zimbabwe.* The Role of Government in Adjusting Economies, Paper 36. University of Birmingham, School of Public Policy.

Sarris, A. and Shams, H. (1991) 'Ghana under structural adjustment: the impact on agriculture and the rural poor'. IFAD Studies in Poverty Alleviation (New York: New York University Press).

Schmidt, G. (1979), 'Effectiveness of maize marketing control in Kenya', Paper prepared for the Seminar on Price and Marketing Controls in Kenya, 26–29 March, IDS, University of Nairobi.

Sen, A. (1981) *Poverty and Famines: an Essay on Entitlement and Deprivation* (Oxford: Clarendon).

Seshamani, V. (1998) 'The impact of market liberalisation on food security in Zambia', *Food Policy*, 539–51 (Elsevier).

Sheng, A. and Tannor, A. A. (1996) *Ghana's financial restructuring 1983–91 in Bank Restructuring Lessons from the 1980s* (ed. A. Sheng) (World Bank: Washington).

Shepherd, A. W. (1979) 'The development of capitalist rice farming in northern Ghana', PhD Dissertation, Cambridge University.

Shepherd, A. and Onumah, G. (1997) 'Liberalised agricultural markets in Ghana: the roles and capacity of government', Paper 12, The Role of Government in Adjusting Economies. University of Birmingham, School of Public Policy.

Smith, M. and Ellis, F. (1997) 'The role of the state in agricultural marketing in Sri Lanka'. Paper 19, The Role of Government in Adjusting Economies, School of Public Policy, University of Birmingham.

Spoor, Max (1994) 'Issues of state and market: from interventionism to deregulation of food markets in Nicaragua', *World Development*, 22: 4, 517–33.

Staatz, John (1992) 'Insitutionalist perspectives on agricultural policy reforms in West Africa', Agricultural Economics Staff Paper no. 92–61, East Lansing, Michigan State University Studies 23: 2, June 1985.

Sukume, C. (1999) 'Quality assurance in liberalised meat markets', in M. Mubbard (ed.), *The Impact of Liberalisation on Public Services to Agricultural Marketing: Case Studies of Change Options in Zimbabwe*. Paper 36, The Role of Government in Adjusting Economies, University of Birmingham, School of Public Policy.

Sung, Bai-Yung (1993) 'Korean agricultural marketing system: its efficiency and perspective', *Journal of Rural Development*, 16: 1, 75–100.

Swaminathan, M. (2000) 'Consumer food subsidies in India: proposals for reform', *Journal of Peasant Studies*, 27: 3, 92–114.

Tawonezvi, H. and Mapika, C. (1995) 'The crisis in the Department of Research and Specialist Services'. Report prepared for the Agricultural Research Council. Final draft (Harare).

TechEcon (Economic and Transport Consultants) 1995 *The National Feeder Roads Rehabilitation and Maintenance Project, the Danida Component* (Copenhagen: Danida Review Report, July).

Technosynthesis (1988), 'NCPB reorganisation study', S.P.A and EEC.

The Economic Review, 29 August–4 September 1994; 2–8 January 1995, Nairobi.

Thompson, A. M. and Lawrence, D. Smith (1991) 'Liberalisation of agricultural markets: an institutional approach', Centre for Development Studies, Occasional Paper no. 8. Glasgow, University of Glasgow.

Trotter, B. (1992) 'Applying price analysis to marketing systems: methods and examples from the Indonesian rice market', Marketing Series no.3 (Chatham: Natural Resources Institute).

Tyagi, D. (1990) Managing India's Food Economy (New Delhi: Sage).

Umali, D., Feder, G. and De Haan, C. (1994) 'Animal health services: finding the balance between private and public devlivery', *World Bank Research Observer*, 9: 1, 71–96.

Umali-Deininger, D. and Deininger, K. (2001) 'Towards greater food security for India's poor: balancing government intervention and private competition', *Agricultural Economics*, 25, 321–35.

UNDP (2000) 'Poverty Report 2000' (New York: United Nations).

UNDP (annual) 'Human Development Report' (New York: United Nations).

USAID/REDSO (1996) 'Report on Kenya Vulnerability Update', July.

Walker, H. and Takavarasha, T. (1997) *Organisational Development in the Public Service: the Case of Self-regulatory Change Teams in the Ministry of Agriculture in Zimbabwe*. Mimeo (Harare: Ministry of Lands and Agriculture).

Walsh, K. (1995) *Public Services and Markets: Competition, Contracting and the New Public Management* (London: Macmillan).

Weeks, J. (2000) 'Central America's free trade flop. Why liberalisation failed to boost agricultural performance', ID 21 Highlights *http://www.id21.org*.

Wolf, M. (1988) *Markets or Governments: Choosing between Imperfect Alternatives* (Cambridge: MIT Press).

World Bank (1981) 'Accelerated development in Subsaharan Africa' (Washington).

World Bank (1992) 'Ghana: agricultural sector adjustment program', Staff Appraisal Report No. P-5523-GH (World Bank, Washington).

World Bank (1993) 'Ghana: agricultural sector investment programme', Staff Appraisal Report No. 12222-GH, Agricultural Operations Division (Washington).

World Bank (1994a) 'Adjustment in Africa: reforms, results and the way ahead' (Washington: World Bank).

World Bank (1994b) 'World development report' (Washington: World Bank).

World Bank (1995) 'Sri Lanka: poverty assessment' (Colombo: World Bank).

World Bank (1996) 'Zimbabwe: fiscal management review' (Washington: World Bank).

World Bank (1999) 'India food grain marketing poliicies: reforming to meet food security needs', Vols 1 and 2. Report No. 18329-IN (Washington: World Bank).

World Bank (2000) 'World development report' (Washington: World Bank).

Zimbabwe Farmers Union (1994) 'Strategies for the Future' (Harare).